Persephone
RISING

AWAKENING THE
HEROINE WITHIN

Also by Carol S. Pearson

The Hero Within: Six Archetypes We Live By

*Awakening the Heroes Within: Twelve Archetypes
to Help Us Find Ourselves and Transform Our World*

*Magic at Work: Camelot, Creative Leadership,
and Everyday Miracles* (with Sharon Seivert)

The Female Hero in American and British Literature
(with Katherine Pope)

*The Hero and the Outlaw: Building Extraordinary Brands
Through the Power of Archetypes* (with Margaret Mark)

*The Transforming Leader: New Approaches to Leadership
for the Twenty-First Century* (editor)

Persephone
RISING

AWAKENING THE
HEROINE WITHIN

Carol S. Pearson

HARPER**ELIXIR**

An Imprint of HarperCollins*Publishers*

HARPER**ELIXIR**

HarperCollins books may be purchased for educational, business, or sales promotional use.
For information, please e-mail the Special Markets Department
at SPsales@harpercollins.com.

HarperCollins website: http://www.harpercollins.com

FIRST EDITION

Designed by Terry McGrath

Library of Congress Cataloging-in-Publication Data
Pearson, Carol
Persephone rising : awakening the heroine within / Carol S. Pearson.
pages cm
Includes bibliographical references.
ISBN 978–0–06–231892–3
1. Archetype (Psychology).   2. Self-actualization (Psychology) in women.
3. Self-realization in women. I. Title.
BF175.5.A72P443 2015
155.3'339—dc23
2015018073

15 16 17 18 19   RRD(H)   10 9 8 7 6 5 4 3 2 1

To Shanna—with love and gratitude

Contents

INTRODUCTION

THE POWER OF STORY IN THE IN-BETWEEN

The Story of Demeter and Persephone *3*

The Eleusinian Mysteries and the Power of Collective Transformation *13*

Why Me? How This Tradition Changed My Life *23*

Why You? Awakening Your Capacities and Potential *33*

Why Now? Thriving in an Unfinished Revolution *43*

PART ONE

DEMETER

Demeter and the Way of the Heart *53*

Demeter Lesson One: Living a Life of Connected Consciousness *65*

Demeter Lesson Two: Demonstrating Brave-Hearted Resilience *77*

Demeter Lesson Three: Valuing the Generous Heart *89*

Demeter Lesson Four: Voting with Your Feet *101*

Demeter Lesson Five: Standing Up for What You Care About *111*

Capstone Exercise: Dialogue with Demeter *119*

PART TWO

ZEUS

Zeus and the Way of Power *123*

Zeus Lesson One: Overcoming the Fear That Fuels a Driven Life *133*

Zeus Lesson Two: Declaring Your Independence *141*

Zeus Lesson Three: Unleashing Your Passion, Focusing Your Actions *151*

Zeus Lesson Four: Regrouping and Rethinking as You Know More *161*

Zeus Lesson Five: Moving from Power Over to Power With *173*

Capstone Exercise: Dialogue with Zeus *184*

PART THREE

PERSEPHONE

Persephone and the Way of Transformation *187*

Persephone Lesson One: Responding to the Call of Eros *201*

Persephone Lesson Two: Claiming Your Love Rights *211*

Persephone Lesson Three: Doing Life a Simpler Way *225*

Persephone Lesson Four: Making Choices to Realize Your Destiny *237*

Persephone Lesson Five: Experiencing Radical Belonging *251*

Capstone Exercise: Dialogue with Persephone *266*

PART FOUR

DIONYSUS

Dionysus and the Gift of Joy *271*

Dionysus Lesson One: Realizing the Eleusinian Promise *287*

Dionysus Lesson Two: Celebrating Life's Great Beauty *301*

Dionysus Lesson Three: Dancing Collective Joy *313*

Dionysus Lesson Four: Directing Your Inner Theater Company *327*

Capstone Exercise: Dialogue with Dionysus *338*

Integrative Capstone Exercise: Your Personal Eleusinian Mandala *338*

CONCLUSION

THE POWER OF STORY
TO TRANSFORM YOUR LIFE *339*

Who's Who *359*

Bibliography *361*

About the Author *367*

Acknowledgments *369*

Notes *375*

Reading Group Discussion Guide *389*

The Power of Story in the In-Between

The Story of Demeter
and Persephone

L ONG AGO AND FAR AWAY, there was an island with an advanced civilization where art and beauty were primary, civility and peace reigned, and men and women lived as equals. But after many ages had passed, people from the nearby mainland—where men ruled and war and violence were common—crossed the sea in their boats and invaded this island, defeating its people. Among the spoils of war they took home with them were some of the island's gods and goddesses.

These gods and goddesses retained knowledge of where they had come from and yearned to return, yet as time passed, their memories of the island and of life as they had once known it grew fainter and fainter. Finally, their recall of the island was like the ghostly writing in a palimpsest, just traces beneath the surface of newer inscriptions, as the deities became more and more what the new mortals they served wanted and understood.

Many ages later, one of these goddesses was luxuriating in the beauty of a verdant field. Her hair was golden, like corn silk, and she moved with the grace of grain blowing in a soft, warm wind. Her eyes were the color of a clear, sky-blue lake. While her body was voluptuous, she also seemed to emerge from the land, with the feel of someone solid and trustworthy. Her very name, Demeter, came from the Greek root word meaning "the mother," and she embodied the compassion and nurturance this name implies.

From his throne high on Mount Olympus, Zeus, the god of all the gods, saw her. Zeus was as muscular as a bull and had the courage of a warrior. Clad from his head to his feet in armor, he inspired the respect of the other gods and terror in the hearts of his people, who knew they must worship him and make sacrifices to him or suffer the consequences. In his role as a sky god, he was known—when crossed—to hurl down lightning bolts, bellow as loud as thunder, and create winds so strong that few could withstand their force.

But on this day, everything had gone his way, and he was feeling secure in his power and happy with his lot. Then he happened to spy Demeter and, filled with lust, he descended from on high to seduce her. Flattered by his attentions, Demeter enjoyed their lovemaking. After they had rested together contentedly, he explained, with some regret, that he had to return to his duties. After all, he was responsible for maintaining the social order of gods and mortals, as well as for quieting the anger of the Titans he had conquered, and soon he would marry the Titan beauty Hera, the goddess of marriage.

Zeus reminded himself that being the chief god, and consequently the king of all, meant that his responsibilities had to supersede his personal happiness, and that Demeter would be fine. Her satisfaction came less from sex or romance than from being a mother, and likely she would gain a child from this union. And he was right in this surmise.

Their resulting daughter was known as Kore, the maiden; it was not yet clear what she would be the goddess of, hence her generic name. She had hair as dark and luminous as the night sky, but a disposition so light and joyful that it seemed as if she had stars dancing as a halo around her. Her skin was honey golden, and her nature was similarly sweet. Her eyes were sea-foam green, the color of the

Mediterranean, and those looking into them often felt a subtle call to adventure that caused them to yearn for something far away and as yet unknown.

Demeter loved her daughter more than anything or anyone, cherishing her and doing everything she could to keep her safe. But one day while Kore was off picking flowers in a meadow with her friends, Demeter left to take care of some business with other goddesses. She returned after a short time, only to learn that Kore was nowhere to be found. Her playmates told Demeter that Kore had wandered off and had not been seen since. Demeter asked everyone in the vicinity if they had seen Kore or knew where she was, but no one would admit to any knowledge of what had happened. Distraught and worried, as any parent would be, Demeter feared that Kore had been killed, raped, or kidnapped. For days, Demeter did not sleep or eat or bathe as she searched frantically for Kore, following ever-widening paths that led her further and further from home.

Finally, Demeter encountered Hekate, the goddess of the crossroads, who was known for the depth of her wisdom, which was especially relevant in times of choice or when someone was at a loss for where to go or what to do. A very ancient goddess, Hekate was one of the few who (along with the Fates) appeared to other gods and to humans in the guise of an old woman. Closely associated with the moon and its phases, she saw better at night, like an owl or a cat. Her hearing, however, was always acute, even catching whispered secrets that traveled to her in the wind. When Hekate recognized the depth of Demeter's maternal grief, her own heart was touched, and she told Demeter that she had heard Kore cry out, and she believed that Kore might have been abducted.

Hekate suggested that she and Demeter visit Apollo, the blazing sun god, who, from his position in the sky, may have seen what transpired. Gratefully, Demeter accompanied Hekate up into the sky to see Apollo. Now, Apollo was a favored son of Zeus, and often served as his emissary. He dutifully explained to Demeter that she need not worry. Kore had become the wife of a prestigious god, Hades, who ruled one of the three major realms of the world. Of course, Demeter well knew that Zeus ruled the sky and the surface of the earth; Poseidon, the seas; and Hades, the Underworld, where the dead reside. But she listened politely, so as not to offend. Apollo went on to assure her that all was well: Hades had asked Zeus for Kore's hand in marriage, and after all, Zeus *was* her father and had the right to decide whom she married.

Hades was a dark and handsome god, rich beyond measure, with a mischievous turn of his lips that women adored. He had loved Kore since he first saw her, but had repressed his growing desire until she was pubescent, and thus of age. When Hades appeared before Zeus, he was lit with passion and trembling with eagerness to hold his beloved. Zeus thought it better to have Hades marry Kore than ravish her unwed, as he feared might happen given what he was seeing. If that were not enough, Zeus knew that Hades had always resented how he, Hades's younger brother, had become chief of the gods, supplanting his older sibling. Zeus had to manage Hades carefully so that he would not stage a rebellion. So all in all, Apollo continued, it was a wise decision for Zeus to bless the marriage then and there.

"So when was this marriage ceremony?" Hekate asked, a bit provocatively, since all the gods should have been invited. Ignoring that question, Apollo explained that Hades had convinced Aphrodite of the depth of his love for Kore and asked her to help him woo her. Aphrodite placed the most beautiful flower anyone had ever seen near the meadow where Kore was playing with her friends. Kore saw this flower in the distance and became so entranced by it that she wandered away from them. She bent over to pick it but found that she had to pull it hard, and when she did so, the earth opened up, and Hades, on his chariot, bounded out of the depths, swept her up in his arms, and carried her back to his underworld kingdom.

Learning this, Demeter feared that, though Kore was alive, she would be scared, upset, even traumatized. Certainly, she was unprepared for sex, especially with someone she did not yet know and who had violated her sovereignty by roughly carrying her off against her will. Kore was a young girl, after all, still really a child, although her body was becoming more womanly. As Demeter ruminated further about this, her worry was matched by her anger at Hades, but even more so at Zeus, who should have protected his innocent young daughter. She did not mind that he had been no help in raising Kore, but for Zeus simply to dispose of her for political expediency and personal advantage was beyond what she could accept.

Demeter was an Olympian goddess and had to stay in her own realm, just as most other gods were confined to theirs. She could not go to the Underworld to rescue her daughter, and defying Zeus's orders was unthinkable. However much Hekate tried to comfort her, there was no comfort to be had. Realizing that the other gods must have known what had happened to Kore but did not tell her out of fear of Zeus, she could not bear to be around them another moment.

Feeling trapped and powerless, Demeter disguised herself as an old peasant woman and set off on a journey with no destination, wandering aimlessly as she fasted and pondered what to do but coming up with no answers. Tired and discouraged, she finally sat down to rest by the sea, in a little town called Eleusis, about fourteen miles from Athens. Kindly daughters of the local royal family saw her and asked her why she was there alone, without family or friends to care for her. She explained that she was from an island paradise but had been captured by pirates who brought her to this place. The young women sympathized with her plight and invited her to their palace, where she was welcomed warmly by the queen, Metaneira, and her attendants. At first, she declined their offers of wine or solid food, but then broke her fast with barley water flavored with mint. For Demeter, the goddess of grain, ingesting the essence of barley worked to remind her who she truly was. Experiencing such kindness from these friendly and welcoming mortals warmed her heart and further restored her hope. Demeter's good spirits were raised enough that she even was able to laugh at the antics of an elderly female servant, Iambe, who did an obscene dance, lifting her skirt to reveal her private parts.

In appreciation for this hospitality, Demeter offered to become a nanny for Queen Metaneira's precious new son, Demophon, a proposal that was accepted enthusiastically, as the queen intuited that there was something remarkable about this visitor, however poor and worn down she might seem. Still disguised, Demeter formulated a secret plan to repay all this kindness by making the son immortal, feeding him ambrosia (the nectar of the gods) and purifying his nature over the fire when he slept. All of which she did for a time, until his mother, the queen, came in at night and saw him in the fire. Of course, she screamed in alarm, yelling that Demeter was killing her son. Outraged that a mortal was chastising her and interfering in a sacred ritual, Demeter erupted. She grabbed the infant prince, threw him down (though he was unharmed), and showed herself in her full goddess glory, demanding that, in order to appease her, the Eleusinians build a temple in her honor.

The ancient stories do not reveal how long it took to create such a temple, but we can imagine that the terrified mortals worked as hard and as long as they could, since gods and goddesses of that time were known for cruelly making mortals pay for any lapses in homage or obedience to their decrees, or even to their whims. What we can surmise is that while wandering, Demeter had no energy to infuse her life force into the crops and other vegetation, which withered as a result of her inattention. Even after she had reclaimed her full identity as a deity, she refused to make things grow. Gradually, a terrible famine that could not be ignored took over the land. Masses of gaunt and starving people appealed to Zeus for help.

Feeling harried and tired from bearing the brunt of all the beseeching and complaining, Zeus called together the Olympian gods, asking each in turn to go to Demeter and beg her to stop the famine and provide the verdant crops she always had before. She was, after all, the goddess who taught humankind the secrets of agriculture. She is softhearted, Zeus explained. She will not want mortals to starve, and she knows, too, that if they stop sending us their sacrifices, we will begin to fade out and disappear. The gods did as they were told, but Demeter remained firm, saying that she would end the famine only when she could see her daughter and know that she was safe and happy.

For the first time in his long reign, Zeus had to come face-to-face with the limits of his power. He was the god of the gods and the chief god for mortals, but he could not make the grain grow. Only

Demeter could do that. Relenting, he sent his son Hermes, the god of communication and one of the few gods able to move between realms, to escort Kore back to her mother. The moment Kore's feet touched the earth, flowers sprung up around her, and in the distance she glimpsed crops beginning to grow again. Anyone viewing her would see that she was striding back to the surface of the earth with a new self-assurance, looking less like a child and more like a young and confident woman.

Kore and Demeter's reunion was warm and sweet, the lovely meadow that surrounded them growing more beautiful and lush every moment they were together, flowers and foliage springing up around them. They embraced, shared their joy, and after a time were joined by the grandmotherly Hekate, leading to more hugs, kisses, and intimate womanly confidences that went on into the evening, and some say continued for days, as women's visits can. Demeter and Kore shared their stories in turn. Demeter described her alienation from the Olympian gods, her wanderings, and how she had been taken in by a kindly royal family. There, she recognized that mortals are not bad, just ignorant, primarily because the gods had failed to educate them. To rectify this, Demeter had decided to create a Mystery tradition to help mortals understand the laws of life and death so they could learn to be happy, prosperous, and free of fear. She would name the rites the Eleusinian Mysteries, after the town of Eleusis, where mortals had come to her aid and built her a temple, and where, Demeter hoped, her daughter would join her in this great work.

Kore shared that, initially, she had been fearful and disoriented after being abducted but drew strength from her mother's teachings about trusting herself and remaining connected to the whole of life so that she would feel at home wherever she might be. And she added that she knew Demeter would be doing everything in her power to find her and ensure her safety. Kore described arriving in the Underworld and how her heart had gone out to the newly dead who did not understand their state. She knew that Demeter would want her to help them. As she did so, their panic abated, and they asked her to become their queen. As queen of the Underworld, Kore explained, she took the ancient name of that far-off goddess, so lately forgotten, who used to occupy that role. She was Persephone now.

Demeter and Hekate committed to calling her Persephone thereafter. Then, with a little laugh, Persephone rather hesitantly told them that Hades had tricked her into eating some pomegranate seeds and then, more proudly, added that she was pregnant, she thought with a god whose gift would be to bring joy to mortals and gods alike. As everyone used to know, powers more ancient than Zeus or Hades decreed that if you eat anything in the Underworld, you have to return there. Demeter and Hekate looked immediately downcast as they realized Persephone's fate, though Persephone continued to show her usual lighthearted spirit, reassuring them that all would be fine. When she was in the Upperworld, she would initiate mortals into her mother's Mysteries, and when in the Underworld, she would initiate the dead into the deeper mysteries that only those who have sloughed off their material forms can know. Hekate, recognizing a need, volunteered to take Persephone's place in the Underworld during the period when Persephone was in the Upperworld. In this way, the dead would not be left bereft. When Persephone descended again, Hekate would return to her role as a seer of the crossroads, helping mortals with difficult life decisions and transitions.

The next part of the story—all that could and can be told to the uninitiated—came from Zeus. Out of gratitude that crops were flourishing once again and sacrifices were wafting up to the Olympian gods, he declared that all should recognize that the seasons of spring, summer, fall, and winter were not accidental but a product of Demeter's will. The seasons remind us of all the cycles in our own lives and that the earth is our mother, and like any mother, she loves her children, her people. From then on, he announced, during the period of winter, when crops are fallow, gods and people alike would take time to honor Demeter's grief over her daughter's sojourn in the Underworld, as well as to honor how this motherly deity will grieve whenever any of her children experience suffering.

He then granted Persephone the right to become one of the alchemical deities who could move at will between the Upperworld and the Underworld. With pride, he also announced that he had invited Demeter and Persephone to rejoin the Olympian gods, and they had accepted. A massive and joyous celebration immediately commenced on Olympus, with mortals below dancing to express their relief and gratitude that the standoff between Demeter and Zeus and the resulting famine finally were over.

Legend has it that after this time, Zeus became a much better ruler—more respectful of the gifts of all the goddesses, less dictatorial and more democratic, less likely to dole out punishments, and more supportive of efforts, like the Mysteries, to help people learn and grow into their better selves, declaring that this story should be told and retold in every generation.

Yet life being as it is, the Olympian gods eventually were overthrown, and Demeter's temple was demolished. Nonetheless, these gods and goddesses still are with us, for those who know how to recognize them. We see their traces in literature, popular film, and human behaviors even today. And discovering how to recognize them in and around you can help you be happier, feel more prosperous, and act with less fear and greater courage—realizing the promise of the ancient Eleusinian Mysteries, the rites that grew up around the narrative you have just read, or, perhaps, truly were created by Demeter and Persephone.

The Eleusinian Mysteries
and the Power of
Collective Transformation

B Y NOW YOU MAY HAVE RECOGNIZED that the island paradise from which the gods came was Crete, and the mainland city to which they were imported, Athens. The historical Athenians actually did practice the Eleusinian Mysteries, in Demeter's temple in Eleusis, which is approximately fourteen miles from Athens. When people participated in the rites, they believed they were at the exact spot where the Mysteries were created, reinforcing the power of the place to deliver on the Eleusinian promise.

What would it have been like to be an ancient Athenian first learning about and then joining in these rites? You almost certainly would have encountered the Mysteries first through a story—similar to the one you just read—of the grain goddess, Demeter: how her daughter was abducted, how a famine ensued because of Demeter's grief, how her daughter returned and then traveled back and forth between the Upperworld and the Underworld, and how this mother-daughter team created the Mysteries. You could be encouraged to identify with Persephone, trusting that the world is a safe place and that, even in death, a cosmic motherly presence is always with you, for she loves you as she loves her daughter, suffers when you do, and works to help you when you are in need. Her story also would affirm that winter always would lead to spring, that famines can be averted and abundance restored, that sorrows will not last forever, and that you can learn the secrets of attaining a happy and prosperous life.

Demeter's choice, to help teach the mortals in the Mysteries instead of punishing them for their ignorance, was radical and unheard of. Before then, when humans offended a god, the gods would punish them as an object lesson. Her decision caused a seismic shift, from a spirituality where mortals were terrified of the gods, and needed continually to propitiate them or else experience excruciatingly cruel punishments, to feeling part of a community of initiates who were cared for by Demeter in life and by Persephone in death. Thus, there was no need to be frightened. The goddesses were more like loving mothers (or fathers) than punitive tyrants. Some scholars believe that initiates were not told exactly what would happen when they died, only that Persephone would care for them then as Demeter did while they lived.

Even without participating in the rites, you would know some things about Persephone, the goddess of spring and renewal and, paradoxically, also the queen of the Underworld, and hence the dead. You would know that Persephone was the priestess of the Mysteries, and that she moved easily between the Upperworld and the Underworld—modeling for you how to move seamlessly from one situation to another and what to do when fate abducts you into a life you would not have chosen—all with a lightness of spirit, great flexibility, and depth of awareness. In addition, you would know a few things about Dionysus, even though he does not even appear in Demeter's story. However, he was prominent in the rites themselves, and his statue was carried in the public Eleusinian procession. You would know he was the god of joy, ecstasy, and dance, who had his own rites as well as being part of the Eleusinian ones. And you would know of the awesome power of Zeus and how Demeter succeeded in defying him.

The first few days of the rites were held in public, so you likely would know what happened during that portion, but even initiates were not permitted to share anything with you about the secret parts (such as, for example, what happened within Kore that transformed her from a frightened adolescent into a confident, mature Persephone, capable of moving between worlds and transforming others as well as herself).

The only way to gain access to the secret lore was to sign up, show up, and go through the nine-day initiation. People from all walks of life were allowed to participate, as long as they understood and spoke Greek (so that they could comprehend what was going on) and had not murdered anyone. If you met these requirements, you would gather in Athens at an appointed time in what is now our month of September. Your first few days would be spent getting briefed and ready for what came next and in celebrations of Athens and sacrifices to various Greek gods. Someone who had gone through the initiation already might mentor you to prepare for the nine-day experience, and/or you might have participated in a shorter, more introductory ceremony held the previous February, which included a purification rite for anyone who had killed someone, similar to the idea of redemption, that sins could be forgiven. These preparations would provide you with important background information.

If you determined that you wanted to become an initiate, as so many did, you would gather with upward of 2,000 people, many of whom would not be the folks you normally associated with, for slaves participated along with kings and queens, and women along with men. If you were a member of the elite, you would be challenged to be peer with people who were your underlings. If you were lower down the status ladder, you likely would feel uncomfortable treating those you ordinarily would defer to as your equals. Like every other participant, you would have brought with you a piglet. (In this populist event, pigs were chosen because anyone could afford one.) At an appointed time, all the pigs would be sacrificed. In today's world, we might imagine that this would mean ritually sacrificing our piggishness (i.e., unbridled greed, lust, gluttony, consumerism, and excess of all kinds). Back then, you and everyone else would eat the flesh; the inedible parts would continue to burn, with the smoke rising as a gift to the gods.[1]

At some point, a cry of "to the sea" would go up, and you and the entire horde would run and jump into the water for a kind of baptism of renewal that was done with high spirits and joy. Over the next couple of days, you would join a procession walking the fourteen-plus miles to Eleusis. Partway through, you would go single file over a bridge, and if you were someone thought to be arrogant and puffed up, hooded figures would shout out, in a jocular tone, embarrassing things that you would rather have kept secret, because no one should go through this initiation without being properly humble. The first night of the procession, you would stay in a temple, where you would sleep, wrapped up like a swaddled baby, anticipating a dream that would provide you with guidance for healing and your next steps in life.

When you arrived in Eleusis, there would be a good bit of milling around, a period of fasting, and dancing with wild abandon through a long night. While at the beginning you might have been a bit awe-struck at being part of such a massive group of initiates, by then you likely would have connected with people and feel supported by them. After all this exertion, you would be exhausted, so there would be a time to rest before the main event, although you might be too excited or anxious to fall asleep immediately.

Sometime the next evening, you would enter the Telesterion, a building large enough to hold all or most of you, for the ultimate transformative experience. You might or might not know that this event would be held during the night, with all these people, in complete darkness, which undoubtedly was scary. You would experience "things shown," "things done," and "things said," which means the ritual included ways of communicating appropriate for visual, kinesthetic, and auditory learners.

Some scholars believe that one or more of the sacred stories would be commemorated in a choreographed dance that everyone did together, and that the birth of a child (most likely Dionysus, but some say Persephone) would be celebrated, visionary experiences would occur, and at some point, a bright light suddenly would appear that would blind and frighten you for a moment until your eyes adjusted. By the time you emerged into the light of day, it may have seemed as if you had died and been reborn, so death no longer held terrors for you. Once the celebrants had come out of the Telesterion, a priestess would hold up two sheaves of barley, libations would be poured onto the earth, and you and the other initiates would gaze up at the sky and cry aloud, "Rain!" and then look down to the earth and cry, "Conceive!" in celebration of the marriage of earth and sky and the fecundity it engendered.

The evidence of history tells us that at the close of this life-changing initiation, you would realize its promise. Even today, investing nine days in order to become happier, more prosperous, and free of fear would not be too shabby. Doing it with so many people also would have instilled an awareness that this powerful experience was not for your good alone. Your personal renewal served the collective renewal of society. Your happiness, prosperity, and freedom from fear supported these outcomes for all.

Before moving on to explore the power of the Eleusinian stories, it is important to provide a bit more historical context. These rites were practiced first by women elders in prehistoric communities who initiated their daughters, and later by men and women of all stations in life. Hugh Bowden, a classics scholar and senior lecturer at King's College London, writes in *Mystery Cults of the Ancient World*, "The Eleusinian Mysteries were the most revered of all ancient mystery cults."[2] Noted art historian Elinor W. Gadon observes that the power of the Mysteries enhanced the prestige of Athens, as "men and women from all over the Mediterranean world came to witness the rites and be party to their secrets."[3]

In *The Once and Future Goddess*, Gadon emphasizes not only the great staying power and regional influence of these myths but also their personal impact and psychological depth, stressing that the Mysteries at Eleusis were practiced "until the fall of the Roman Empire.

Men and women, philosophers and kings, came from all over the known world to be initiated into her [Demeter's] mysteries. This placed the neophyte into direct and personal relation with the sacred. Well established by the seventh or sixth century BCE, the Eleusinian Mysteries for a thousand years were at the center of inner religious life." Gadon continues, "Classic literature is full of ecstatic accounts of the initiates' transformation at Eleusis. Homer tells us, 'Happy is he among men upon earth who has seen these mysteries,'" and Sophocles credits the Mysteries with furthering happiness in individuals and the community.[4]

The Mysteries attracted men and women from far and wide and were a major influential force in Athens when it was incubating, and then birthing, democracy, drama, philosophy, and empirically based science—all of which have been crucial to the development of Western thought and to attitudes and behaviors we take for granted today. Much about ancient Greece has been mined as part of our cultural legacy, and many of us learned in school about Greek philosophers, such as Socrates, Plato, and Aristotle; scientists, such as Pythagoras; and dramatists, such as Sophocles—or we encountered the Greek myths, including the story of Demeter and Persephone, in Edith Hamilton's *Mythology* or another compilation. However, it is unlikely that you have heard about the Eleusinian Mysteries before now unless you happen to be a myth scholar. At best, most people today know only the rudiments of the story of Demeter and Persephone and regard it as describing rather quaintly why we have the seasons of the year. It is an important missing piece of our collective heritage.

Archaeological evidence suggests that the rites were practiced in some form at least beginning in the fifth century BCE, but some sources say from much earlier than that, and historical evidence indicates that invading Goths destroyed Demeter's temple around 395 CE. The story that begins this book is based loosely on the *Homeric Hymn to Demeter*, which is considered the most authoritative version and written sometime between 650 and 550 BCE.[5] The practice of the Mysteries was at its height at the time ancient Athenians were most powerfully developing the aforementioned innovations: democracy (fifth through third

centuries BCE), philosophy (fourth through third centuries BCE), medicine (fourth through third centuries BCE), and empirical science (third
century BCE). The Mysteries may have influenced and been influenced
by all these developments.

Manly Hall, scholar and founder of the Philosophical Research
Society in Los Angeles, reported (in *The Secret Teachings of All Ages*)
on research that demonstrated how the teachings of the Eleusinian
Mysteries have come down to us through many esoteric metaphysical traditions and have remained crucial to Freemasonry, which kept
alive a spiritual interpretation of the Demeter and Persephone story
along with the dream of democracy. Because most of the founders of
the United States were Masons, as was Hall, he suggests that there is
a direct relationship between the inclusiveness of these Mysteries, as
well as their dream of a more egalitarian world, and the formation of
America's democracy.

The reference in my version of the Demeter and Persephone
myth to Crete was not in Homer's poem, but it is historical. Demeter, Persephone, Zeus, and Dionysus all were originally Cretan gods.
It is likely that when Demeter says that pirates abducted her from
Crete, Athenians would know that she was referring to the Greek
takeover of Crete and many of its gods, including those celebrated in
the Eleusinian Mysteries. Whether or not people believed that these
times literally existed, the history was preserved in the stories that
infused their culture with its meaning. The ideal of Crete served as
the mythic or historical Eden in the Mysteries, having much in common with our modern utopian dream of attaining sustainable peace,
social justice, liberty, and the right to pursue our own happiness in
our own way.

Riane Eisler, in her groundbreaking book *The Chalice and the Blade:
Our History, Our Future,* makes a compelling case that prior to the
establishment of patriarchy, partnership societies did exist where
women and men were respected equally and that Minoan Crete was
a place that preserved such a society in a culturally advanced, prosperous civilization—although the partnership culture was wiped out,
likely by Greek invaders, sometime between the fifteenth and elev-

enth centuries BCE, a period corresponding to the earliest records of the Eleusinian Mysteries.

Even as late as the classical period in Greek history (480–323 BCE), Greek mythology retained the memory of prepatriarchal times and the establishment of patriarchy. For example, the myth about the Temple of Apollo at Delphi, which is credited with helping to civilize Greece, unapologetically recounts how originally this was a temple dedicated to the goddess Gaia, the Earth Mother, and guarded by the earth-dragon Python, which was depicted as a serpent, an ancient symbol for regeneration (as in snakes shedding their skins). The Greek god Apollo defeated Python in battle and conquered the temple as well as the priestesses who were its oracles. The precursors to today's Olympic Games were held near this temple, which was part of a larger site devoted to Apollo. Winners were honored with a laurel crown, cut from a tree by a boy who reenacted Apollo's slaying of Python.

Although the Mysteries primarily attracted individuals who wanted to be happier and more successful, they also had a less emphasized social function. At the time the Eleusinian Mysteries were at their height, women had many ideas about how to improve society and their roles. The clever women and family groups who evolved the Mysteries complemented (some would say undermined) the Athenian role-defined hierarchical society by opening their rites to everyone—men, women, kings and other elites, concubines, slaves, and so on—and by the woman-honoring nature of their teachings. With this in mind, the rites were designed very much as a collective event, forming a wide network of individuals joined together through a common experience to support egalitarian values. In some ways, the Eleusinian initiation prefigured modern social-change strategies—the women's movement of the 1970s, the civil rights movement, the gay rights movement, the environmental movement, the peace movement, and the new age movement—while also being more celebratory and enjoyable than most. We also can see it as the forerunner of today's self-help movements, promising initiates that they will gain the secrets of living a better life.

The Eleusinian Mysteries were passed on orally and through initiatory experiences but had no sacred text or writings. This might have occurred because not everyone involved could read and write. Nevertheless, it had the positive impact of allowing the tradition to evolve naturally to meet the needs of new times. By focusing on Demeter's desire that mortals learn through experience to mature and develop, the tradition avoided dogma and was not rule based. It contained no set precepts about what you had to do or could not do (except murder). Thus, its teachings could change as people did.

This also protected the rites from censorship, since there were no publicly described beliefs for anyone to critique or ban. We now know about the Mysteries through various literary sources and through side comments by famous philosophers and others who referenced the importance of the rites (and often their own participation in them). From these, we learn that you would go through the rites once, or perhaps twice, as a deep-dive initiation, but after that, there were no required practices or services to attend.

Because the deeper secrets of the Mysteries never were written down and initiates were forbidden to share them—and historians say they never did so—we do not know for sure what happened. Moreover, virtually all of the accounts are by prominent men, who might have interpreted the Mysteries through the lenses of their own perceptions about gender and about women. Fortunately, anthropologists have traced material remains from the initiation site, scholars have linked together references to the Mysteries in major literature and philosophical writings of that era, and the Eleusinian stories are so archetypal that they recur in poetry and prose fiction into our own time. From these sources, I have attempted to decode enough about what the secrets disclosed in the Mysteries likely were, in order to provide you with a virtual initiation through reading this book and doing the exercises it contains.

The Eleusinian stories, when upgraded for the modern age, offer models and practices that can help any of us find, and be true to, ourselves while living in a world in transition. For all these reasons, I've written this book for women like you—heroines—who want to maximize their potential, not by following some prefabricated script for what they should be, but instead by discovering and following their true desires, living according to their authentic values, developing their innate strengths, and feeling fully alive in the moment, sustained by an undercurrent of quiet joy. It also is written for men who want to understand and empower the women in their lives—sisters, wives, daughters, friends, colleagues, employees, clients, customers, and so on—and especially for the large number of men who already are committed to being equal partners with women, at home and at work. It can be difficult to negotiate the bumps in the road in the absence of clear norms for how to do this or of structures of social support.

Because I believe that no one should be teaching others what they have not learned themselves, the next section of this chapter begins on a personal note, explaining how the story of Demeter and Persephone grabbed my attention and eventually motivated me to write this book. This section also contains a brief discussion of archetypal characters and mythic stories and how to use this knowledge to understand the Eleusinian Mysteries stories and apply them to your life.

Why Me? How This Tradition
Changed My Life

I N MANY INDIGENOUS TRADITIONS, a person seeking answers
to questions would approach a medicine man or woman sitting
by the fire and ask what they should do to resolve their dilemma.
He or she classically would respond to this request by saying, "Let me
tell you a story." Moms, dads, mentors, and friends can do this, too.
Telling a story that was just right for the person, the situation, or the
time required great psychological understanding, empathy, and narrative intelligence. In new situations, often what is needed is a new story
that can help supply a map for a new journey and a toehold when that
journey feels like climbing up a steep and dangerous mountain.

Today, many depth psychologists respond to clients similarly, connecting their suffering to a mythic story that provides perspective.
Often, the narratives they choose are about ancient gods or goddesses
from a variety of cultural traditions. The analyst James Hillman, for
instance, believed that our pathologies were ways that the gods were
getting our attention, and by "gods" he meant "archetypes," a belief
he developed in his seminal book *Re-Visioning Psychology*. Psychiatrist

C. G. Jung coined the term "archetype" to mean a psychological pattern that, because it is universal, includes its expressions in all people, times, and places. When one is awakened in you, the part that is relevant to you and your path forms a subpersonality (an aspect of your psyche that expresses that archetype's motivation, affect, strengths, and narrative). If you call upon an archetype, you are not asking some external being to come into you. Rather, you are calling on a dormant strength that has always been within you to wake up and be available to you—which is not that different from beginning a new exercise regimen and then noticing muscles you never knew you had. Archetypes can be seen in art, literature, human dreams and fantasies, and human creations. So linking our individual challenges and troubles to an archetypal story is a way we can connect with patterns that are universal in human consciousness. Where do these come from? Jung said that they emerged from the collective unconscious of the species. Others have postulated that they may be in our actual DNA.[1]

Sometimes, even without such a helpful guide, we happen upon a story that resonates so strongly with us that it stays in our consciousness for a long time and guides us in some way. You may have had this happen with a sacred text, a book of fiction or a biography, a movie, or even an anecdote someone happened to share with you. The story of Demeter and Persephone has had that role in my life, and I share some of my personal history as a way to illustrate how such archetypal and mythic stories can serve as a guide on your journey to greater happiness, abundance, and freedom.

The story of Demeter and Persephone grabbed my attention at a time I had been feeling at the top of the world but then had an experience that left me reeling. When I was in my late twenties, I was thrilled to obtain a faculty job at the University of Colorado. Luckily, I had received my Ph.D. just as the academic job market was opening up for women. At the university, I was assigned to a to-die-for office with sliding doors leading out to a balcony overlooking the Flatirons, the dramatic outcroppings at the foot of the Rocky Mountains, a setting that was my idea of heaven. The office also was inhabited by a man so attractive that the first time I saw him, I had a premonition that I needed to get out of there—danger lurked! Of course, I soon remembered that I was a serious professional and needed to act like one: in this case, to work in my assigned space. So I rearranged the furniture in that big office to block my view of him and settled in.

I was on a roll. I lived in a wonderful place, had my dream job, and within a few months began an enticing friendship with just the man I had sought to avoid: my officemate, David. The father of three young boys, David was recently divorced from his wife, who was spending the year abroad, and he was raising his three young sons by himself. They were darling and rambunctious. Soon we began to date.

At that time, Boulder, Colorado, was a hotbed of new thinking that was questioning racial attitudes, traditional gender roles, the war in Vietnam, and the military-industrial complex that President Eisenhower had warned us about. Drugs were abundant, although I was not into that, but reports of what some people were experiencing on LSD raised questions about the very nature of what was real. I was branching out in so many ways, including by dating someone from New York City who came from a Jewish background, and expanding my ideas, many of which had developed while I was growing up in Houston and attending the Presbyterian church. Revolutionary thinking was all around me, and I felt as if I was experiencing my own personal version of a revolution, where everything seemed possible.

But sometimes fate also tosses us into a story we would not have chosen. One day, after David and I had been dating long enough that I had become close to his children, he and I were in our office working when word came that Douglas, David's youngest son, had gone missing while in the care of a woman who had always been a reliable babysitter. It was one of those moments when the bottom just drops out of your life. Later that evening, his body was found, and we learned that he had drowned in a nearby creek that usually was very shallow, but from the spring runoff was then deep and raging. I was undone by grief for the loss of a child I already had come to love, and also by the shock at how life can change in an instant. I went from feeling on top of the world to being totally dejected. I learned there is no golden highway to avoid life's traumas, tragedies, losses, or even unnerving disappointments. As any father would be, David was devastated. And of course, his other sons were thrown into a level of traumatic experience they were too young to fully comprehend, just at the time they were grasping the fact that their parents' marriage had ended. But David is a strong man, and he mobilized to focus on doing what was needed to care for his children and continue his work.

Over the next few years, David and I moved in together (with the boys), married, relocated from Colorado to Maryland, and had a daughter. At the beginning, I was the least hurting person in my new family, so I appropriately concentrated on supporting David and the boys in their grief and pursuing my career. However, I remained rather disoriented and full of questions—about life and loss and how to make the best of a difficult situation. I needed a story that was more complex than "Have faith, God works in mysterious ways," or "Think positive, and happiness will be yours," as helpful as any of these might be.

I needed a deeper story to guide me through.

During this disorienting period, I happened to remember a Greek myth—yes, the one this book is about—wherein the goddess Demeter finds that her adolescent daughter, Persephone, suddenly has disappeared. I completely identified with Demeter's suffering because of the anxiety and fear David and I felt when we were told that Doug was missing, and the deep grief we, and Doug's mother, suffered when his body was found and we understood the finality of what had happened.

Somehow, in remembering the story of Demeter and Persephone, I felt comforted by imaginatively identifying with Demeter, who had an experience analogous to mine, though Persephone's return to her mother did not register that strongly in my mind. The fact that she was in the Underworld meant to me that she was dead. The existence of this story thus helped me feel solidarity with others, including my own dear mother, who have endured such losses, enabling me to recognize that loss is part of the human condition and my family legacy. While I was growing up, my mother endured two miscarriages and two stillborn twins, the death of a brother to a heart attack, and the sudden drowning death of a sister.

Recognizing archetypal patterns in stories also can help stabilize us in a crisis because they tell us what fundamental human narrative we are living and what it requires of us, so we have a clue what to do. The experience of the loss of someone you care about is a typical challenge in a love story, which is why many of us are hesitant to love. We fear that we cannot survive the loss of someone we treasure so much.

Sometimes a family legacy makes it easier to recognize the deeper story when we are in it. My parents taught me to love and to show that love to my neighbors by helping when they are in need, to show love for my children by caring for them, to show love for my life partner by being there for him through thick and thin, and to recognize that love is a window to the soul that links us to what is best in us (which they called the inner voice of God). English has only one word for love, but the Greeks teased out many more; however, I did not need to use fancy Greek words to recognize the many forms of love important to my life. Reading Joseph Campbell in graduate school taught me to "follow my bliss," which would help me realize my heroic path, and then the women's movement added the understanding that I might also love myself. So I already had language and concepts to help me understand and experience six dimensions of love, several of which are illustrated in the story of Demeter and Persephone.

When we speak of "love" or "loving," we are referring to a concept or process that can seem rather abstract. Demeter and Persephone embody archetypal elements related to loving. Telling a story or creating an artistic rendering of Demeter or Persephone brings a particular expression of love to life, so that it becomes real to us and provides an imaginative encounter that makes it easier to find that quality in ourselves, recognize it in our past and present experience, and see how it might serve our future wholeness.

Looking back over my life, I realized that although I identified with Demeter's worry and loss, I also had been living Persephone's story at the same time, without recognizing it. Falling in love with David abducted me into a love story, and the challenges of being a wife, stepmother, mother, and professor with students and colleagues who counted on me forced me to "put on my big girl panties" and grow up. If I were to be true to my love for David and his children, I needed to shed my youthful narcissism, to learn to be there—for him and for the boys—not just when it was easy but also when it was hard. From this, I learned how archetypal stories also help us mature as humans, by providing plotlines that guide us.

Those first few years together with Jeffrey, Stephen, and then our daughter, Shanna, were in some ways an exciting time for David and me—in our careers, in our passionate love for one another, and in forging a new family. In other respects, however, there was always a backdrop of sorrow. Although Demeter succeeded in getting Persephone back, *we* had to accept the finality of Doug's death. The challenge of the Demeter archetype requires any of us, but particularly those who have experienced loss, to resist pulling back in fear from trusting the love we have for one another and instead to love more fiercely and fully. To the degree that we have the courage to do so even most of the time, we find the treasure of a happy marriage and a happy family. David, these amazing boys, and our equally amazing daughter are my bliss, as are my darling grandchildren and the work I have been privileged to do.

Archetypes that arise in us are allies to our vocational callings, since the need to work is as primal as that of love. Humans have to work to survive—at least many of us do—and prosperity is furthered by more and more of us doing work that is right for our skills, interests, and motivations. But deeper than that is a work instinct energized by archetypes. When I was a little girl, people would give me dolls to play with, and I would line them up and teach them, so it seemed clear that my future should be in education.

By the time my daughter was born, I was charged with directing a women's studies program in addition to being a professor, a responsibility that awakened in me a dormant archetype, in this case something I was called to do rather than one that I knew I would do. In that role and in my subsequent administrative positions, I've always been a change agent, bringing new ideas into organizations.

That means I had to deal with the backlash when members of the old guard, who often were Zeus-like, felt threatened. Thus, I'm personally very familiar with the Eleusinian archetypal drama of Demeter standing up to Zeus to protect who and what she cared about. My efforts always felt to me as if they were done with grace and in the interest of care and inclusion, but I suspect that is not how they looked to those who liked things as they had been. Actually, I know they did not. To succeed in my work, I needed to integrate Zeus into my psyche, rather than be at war with him. And when, like many women today, I became driven, and thus exhausted, it was Dionysus who came to the rescue, encouraging me to learn to dance.

My most important writing has been motivated by the inner Persephone part of me that always yearns for more wisdom about the interior life of people and social groups and about how knowing ourselves also furthers an ability to be intimate and authentic. I can see that her role in my life was nurtured by my coming from such a devoutly spiritual family that encouraged me to pray about my problems and receive guidance, and later by my study of Jungian psychology and other depth psychologies in graduate school and postgraduate studies. It also was the Persephone archetype in me that motivated me to engage in pursuing education about ecumenical psycho-spiritual traditions, which led me in midlife to enroll in a doctorate of ministry program and earn a D.Min. degree. And although it was the call of Demeter that focused my attention on the Eleusinian stories, it was Persephone who demanded that I write this book. I'm not prone to hearing voices, but some years ago, when Shanna and I were exploring the ancient Minoan ruins in Crete (at Knossos, which dates from the time the gods and goddesses in this book were known in their earliest forms), I actually felt as if a voice inside said, "Bring back the Mysteries." (And yes, of course, I know that that voice could be just from my own imagination, which does not make it any less of a genuine calling.)

Recognizing over time how useful these archetypal energies could be to others besides myself, I have engaged in greater in-depth research on the story, and then on the mystery tradition to which it gave rise. In that process, I've found that many other women and men who know of it relate to this story as strongly as I have, even when the details of their losses, challenges, and desires are different than mine. This research, too, has provided context for recognizing the interdependence of my life with patterns of historical change in recent times that parallel those recounted symbolically in the Eleusinian stories and rites, where women similarly challenged the powers that be. Had opportunities not opened up for women when they did, it is unlikely that I would have led institutes and a graduate school, been a published author, and, at the same time, been a wife and mother. From this I realized that archetypal plotlines that emerge in our lives are interrelated with those active in the larger society around us. The larger patterns affect us, and our choices also affect them.

My challenge in writing this book has been to be true to what the Eleusinian Mysteries stories meant in the context of their times while also highlighting their relevance to women today. The version of the myth of Demeter and Persephone that begins this chapter uses Homer's telling of the story as the basic text, but I have altered it in small ways that seemed likely to be truer to what women would have been teaching than to what men of their time were writing down. To prepare for writing my version and this book, I steeped myself in numerous ancient accounts of the story and in what modern scholars have postulated about them and about the secret parts of the rites. I also used my background in narrative and human psychology to recognize parts of the plot in ancient and modern sources that simply did not make sense within the larger narrative.

In doing so, I put myself in the mind-set of a judge hearing testimony, being closely attentive to internal contradictions and conjunctures that are at odds with known facts. I also kept in mind that women originally created these rites. I then meditated upon the story, knowing that the filter in my own mind would reflect a modern woman's perspective that naturally would highlight its relevance to our time, though ideally without falsifying history. To supplement this, I paid attention to how the deeper archetypes within the stories have been visible in our times, and have woven these manifestations into the discussion of each of the gods' and goddesses' stories.

As you read further in this book, I invite you to sort out archetypes within you that are family legacies, allies to support your individual gifts, or ones needed only in response to a particular situation. More generically, their stories also confront any of us with questions that help us find ourselves and our purpose. Demeter asks: "Who and what do you care about, and what does that require of you?" Zeus: "What are your talents and strengths, and how do you need to develop them to make your way in the world?" Dionysus: "What do you love to do, and what genuinely interests you?" And Persephone: "What does your intuitive sense of guidance tell you about whether this or that path or person is right or wrong for you?"

Their stories also can help you find the answers to these questions and then act effectively on what you have learned. These teaching stories, which provide the meat for parts 1 through 4, illustrate how you can develop the gifts of these archetypal gods and goddesses, avoid being derailed by their counterproductive temptations, and get back on a track that is positive and authentic for you if you find yourself expressing their less desirable aspects. If you apply these lessons to your life, the result may well leave you with the joyous feeling of being like a puzzle piece dropped into its right place where you can be happy, experience abundance, and be (relatively, at least) free of fear.

To facilitate this outcome, the following section contextualizes all this by exploring the power of myth to express an archetype in a cultural framework, so that it is particularly helpful in a specific time and place, and also the primal nature of the four Eleusinian archetypes and how they arise out of the intersection of human physiology, psychology, and the nature of life on this earth.

Why You? Awakening Your Capacities and Potential

ARCHETYPES THAT ILLUSTRATE human ways of being and aspects of our experience are deeper than culture, but their particular expression in images, characters, and narratives inevitably reflects cultural mind-sets and specific situations. So while the most universal form of the archetype persists over time, its expression evolves and regresses, depending on the consciousness of any given culture at a particular moment, which also may vary by subgroups and individuals. Thus, the Eleusinian archetypal stories mirror the consciousness of ancient Greece, but the deeper universal truths that shine out from beneath the cultural trappings are seen in my life and likely in yours. The way we experience them necessarily will reflect our own time and the subculture and larger culture in which we live.

Modern understandings of ecology add further perspectives on the power of archetypal stories in human life, which seem to me particularly important in a book that begins with the story of Demeter as Earth Mother. Ecologist Duane Elgin concludes from his study of natural and human evolution that our makeup replicates patterns in the larger universe, so learning about this massive context for our lives helps us with our own growth and development: "Our universe appears to be a living organism of immensely intelligent design[1] that is able to perform multiple functions simultaneously," he writes. "The same dimensional geometry that structures physical reality also structures our perceptions, creating the environments that both people and civilizations move through in a learning process. Although your universe comes with no explicit 'operating manual,' you do not need one, as the fabric of reality has embedded within it the evolutionary insights that you seek."[2] The fact that we are made from stardust can save us from feeling separate from the larger universe, however awesome and immense it appears. Archetypes provide structure to our perceptions about physical reality, and human narratives about them provide a kind of operating manual that can help us feel more at home in the world as we also become more authentically ourselves.

There are many archetypes that motivate us and guide our paths, but I have come to believe that the special power of these four (Demeter, Zeus, Persephone, and Dionysus) derives from how primal they are, helping us connect with the deeper challenges of being human.

The shallow, soulless stories that predominate in our materialistic society (and imply that the secret to happiness is to acquire money, status, and things) have led to much of the ennui of modern life. To restore the ability to live fully, we need to get back to basics. Our evolutionary path is embedded in our bodies, and then augmented by archetypal stories that help us realize our potential to live full and vital lives. If you ever feel like you are living on the surface of life and hungering for something more real and authentic, these archetypes can help you return to the human basics that can fulfill that longing. As Elgin's formulation suggests, our primal nature and our evolutionary trajectory are built into our physiology, in dynamic interaction with conditions of life on Earth. For example, to survive, we need water, food, clothing, and shelter, and to learn ways to put off our inevitable mortality for as long as we can. A good many of us need to reproduce for us to survive as a species. And our biology predetermines so much else, too.

Humans are born with big brains, so obviously we are supposed to think. This includes understanding logical consequences (if we are not careful, we can get ill, provoke retaliatory violence, alienate the people we count on, etc.). With these brains and their creative potential as well as our opposable thumbs, we are designed to invent things and use our intelligence to notice whether or not they make the world better, and if not, reinvent. We also are programmed to understand the logical consequences of our behavior. For example, if we use up the food supply in our region, we will need to move on to what might be more hostile territory or else starve, so best to prevent the famine in the first place.

Because of our large heads, we have to be born before we are developed enough to survive without extensive and ongoing care by our parents, which means that we need to learn to please them, which in turn requires us to develop the capacity to track the stories in their heads that are guiding their actions—that is, to extrapolate patterns of meaning from the behaviors of others. Moreover, our minds crave meaning and get rattled when they cannot figure out why things are happening and what to do about them, leading to our thinking in narrative forms: "Something is happening, and I need to do something. What is that something? What is likely to happen if I do that? And then what?" And on and on. It is out of our early meaning-making that an archetypal story begins to guide our actions, as Demeter's story did for me, given the caring values and priorities of my parents.

But then we grow up and have to meet our own basic survival needs, finding food, shelter, and fulfillment, which means we have to work, formulate and achieve goals, and understand context. In hunting and gathering times, people would have needed to know where the edible animals and plants hung out, and after agriculture was invented, to understand what crops would and would not grow where. Because it is very challenging to survive alone, much less fulfill our potential, we also need to get along with others and be useful to the group.

In earlier times, Zeus and Demeter split up these tasks. When they did so in ways delineated by sex roles, the human mind began to associate the ability to perform certain tasks with gender and to see the archetypal configurations that allow us to succeed in those tasks in masculine or feminine forms.

Unlike most other primates (with the exception of the bonobo), humans have sexual urges much of the time, not only during a mating season. This potential for frequent sexual activity can lead to a great many offspring that have to be cared for over many years, so our biology sets us up to learn to care for others. Humans are herd animals and survive best when we care not only for children but also for one another; thus, altruism also is important for survival. In early human communal life, everyone would have needed to be involved in protecting, nurturing, and educating children and likely all other major activities related to group survival, even though men might take the lead in some and women in others.

Once agriculture was introduced, which the Eleusinian Mysteries credited to Demeter, people had to learn to care for and partner with the land. This cluster of activities helps make sense of how Demeter is associated with the earth, mothering, altruism, collaborative group functioning as needed for gathering, and all the elements of caring for one another, including food preparation and domestic and other crafts that created a sense of home. Today, the Demeter archetype helps us to feel fulfilled and that our lives are meaningful as we do these tasks, and to hold out against society's tendency to undervalue the importance of compassion, kindness, and generosity. We see Demeter's influence in the nonprofit world and other charities, the environmental and peace movements, and the domestic sphere.

Because men on average were physically stronger than women and not hampered by pregnancies and nursing, they were responsible for hunting and warding off invaders, preventing violence through negotiation, and helping to organize larger social gatherings where smaller groups from the same region came together. As part of that, they also may have taken the lead in bartering and trading goods as they traveled about. They naturally took responsibility for making weapons and other tools related to their work and for building projects that required physical strength and uninterrupted work time.

Today, the Zeus archetype helps us feel alive and vital when we are forming hierarchical teams, making lists and completing tasks, avoiding and engaging in conflict and competition, managing power relationships between potentially competing groups, and protecting whom and what we love. His priorities are reflected in business, politics, the military, much of rule-based organized religion, and traditional management theory and practice.

Sex, in human life, is not just for reproduction. We also are made so that we are capable of having intercourse while looking into someone else's eyes and having our hearts pressed together, not to mention enjoying deep, soulful, lingering kisses. We clearly, then, are supposed to seek romance and intimacy, with some of us making a commitment to long-term mates. Such coupling also satisfies a deep human need for intimacy over an extended period and encourages men to stay involved with their children.

Because tending to children takes a good deal of time, early societies needed some people who were not tied down with such family responsibilities. Nature guarantees that a certain percentage of us will be gays and lesbians, perhaps to ensure that not all sex will lead to offspring. Our biology shuts off reproduction at a certain point in a woman's life, freeing her up for other tasks. Nature also reduces women's capacity for reproduction in times of famine, but even ancients knew that there could be better ways than starvation to accomplish this. Those who had gifts as medicine women or shamans for a tribe would have many time-consuming responsibilities for health and healing and rituals, so they possibly may have chosen a life with few or no children.

Persephone's concerns seem to cluster around the many aspects related to becoming conscious enough for free will to kick in. This would include romantic love, where commitment becomes an act of choice, self-awareness and self-discovery, the selection of a vocation, and your own connection to a numinous, unseen world. Persephone is a goddess that supports free will, as she herself makes unconventional choices like living with her husband half the year and near her mother the other half and becoming the initiator of the living and the

dead. Today, Persephone helps us feel alive and solid in our identities as we find ourselves, show up in intimate romantic and friendship
relationships, and seek deeper meaning for our lives. And we see her
influence in love stories, the self-help industry, spiritual movements,
and the new age movement.

In indigenous groups, people unencumbered by children often
become the fools, jesters, and tricksters who liven things up, so we
can extrapolate that they may have done so in humanity's early days.
If we watch young children, we can see that it is natural for us to be
playful and also to be rebellious and to break rules—evidence that free
will is embedded in our natures. In addition, our bodies are very flexible compared with those of many other species, allowing us to play,
dance, and experience primal vitality in ecstatic movement and physically demanding sports. Because humans have the gifts of creativity
and humor, we also create and enjoy art and laughing together. Dionysus as an archetype furthers activities that make us happy individually but is even more important to those that allow us to experience
collective joy.

Today, the Dionysus archetype helps us move out of our driven
lifestyles to relax and enjoy ourselves. This includes finding ways to
love all the parts of our lives: family and home, our work, our quests,
even our yearnings, and it provides the opportunity to be playful in
whatever we are doing rather than just being harried by all the tasks
on our to-do lists. In the Eleusinian tradition, he is not portrayed as
a character in the same way that Demeter, Zeus, and Persephone are.
Rather, he is the embodiment of the happiness attained at the end of
their stories, adding the kind of joy a new baby can bring to a family.
His chapters thus serve as a summary that shows how all four archetypes complement one another.

There is a prototype for his linking the gifts of these gods in our
human physical functioning, as he is a god who celebrates feeling good.
We are preprogrammed with chemicals in our bodies—dopamine,
oxytocin, and serotonin—that reward desirable behaviors with pleasure and nag us when we do not engage in them with a feeling of an
inner lack of something. It is even possible to get addicted to one or

more of them if you do not have a good balance between them. In *Meet Your Happy Chemicals,* Loretta Graziano Breuning discusses how these chemicals triggered by our brains reinforce behaviors, emphasizing how they further individual and group survival, and also how they promote our ability to thrive. However, such chemicals typically are produced in spurts—and once they recede, we go back to neutral again, which encourages us to keep doing what we were doing so that we will continue to produce them.

From Breuning's description of the behaviors the chemicals reinforce, it is easy to see what archetypal gifts and stories they also support. Dopamine reinforces having goals and the process of attaining them, and thus is the Zeus inner chemical of choice. Oxytocin strengthens social bonds through caring and love in all its forms, so it is the reward we get for living Demeter's and Persephone's stories and for dancing together with others, as Dionysus wants us to do. Serotonin is released when we are doing things that gain us respect and status and other rewards that allow us to feel like safe and valued members of the human community. Any of the archetypes can help us do this, for all contribute to the health and wholeness of a group. Many of the most popular antidepressants on the market today are cures for a serotonin deficit. We know that women use such drugs more frequently than men, but we do not know if those who feel they are of lesser status than others also are prone to shoring up serotonin by taking something external. Perhaps mirroring one another more positively or working on our own self-esteem by making a contribution we value and feel good about might increase internal serotonin. It is worth a try to use archetypal therapy to encourage our bodies to produce chemicals that make us happy by engaging in the activities they reinforce.

A different category of chemical also has a role in alleviating pain. Breuning notes that the primary function of endorphins is to dull physical and emotional pain for a short time, supporting the notion that nature has a compassionate side, just as Demeter's story tells us it does.

The knowledge that our physical natures also positively reinforce what we should be doing through pleasure and negatively reinforce what we should not be doing with pain can be part of a heroine's playbook, allowing her to recognize at the feeling level when her body is saying yes, that she is on the right path, or when to pay attention, as danger lurks and she might need to change direction. Adrenaline generates the fight, flight, or freeze response for quick action to escape harm. New research has found that adrenaline also encourages the "tend and befriend" capacity in women, which is very helpful if you are, say, a relief worker in an evacuation center with waters rising or in a refugee camp with no immediate prospect of more help being on the way.

Unpleasant chemicals give us a heads-up that there is a less urgent problem we should be paying attention to—most often cortisol, which in small amounts causes stress, and in large amounts, fear. For survival reasons, these unpleasant feelings last longer in the body than do those that make us happy, and failure to deal with what is causing the cortisol to be released results in more and more unhappiness, plus feeling stressed out and gaining weight, particularly around the midsection. Our bodies keep producing these chemicals until we face the origin of the problem.

In all these ways and more, the Eleusinian Mysteries narratives can help you awaken and develop aspects of your psyche essential to the heroine's path. Most of us are trying to make our way through life without utilizing all the capacities we were born with and wondering about why things do not work out well. This is just like having all sorts of software programs on your laptop that you never use, even though they would help you have greater ease in what you are trying to accomplish. Our egos are there to help us survive; our hearts to care for one another; our deeper selves, or souls, to yearn for meaning and to reach out to experience things beyond mere survival, like true love; and our bodies to enjoy the pleasures of life. Demeter's story offers guidance in opening the heart, Zeus's in developing healthy ego strength, Persephone's in connecting to your deeper self (or soul), and Dionysus's in tapping into the wisdom of your body.

Living these four stories, even vicariously by reading them, acti-

vates these inner psychological resources, providing you with a simu-
lated initiation. And each of their stories also supports your ongoing
growth and development, wherein these capacities in turn develop a
better connection with the power of the mind, with the result that
your capacity to think and make decisions matures. Their stories also
help with the how-to of utilizing these capacities so that they are
engaged and enable you to avoid the paralysis that strikes when the
ego, heart, soul, and body are tugging at you from so many different
directions that you cannot find a way to move forward.

The final section of this introductory chapter differentiates myths
and archetypes and describes how understanding archetypal narratives can provide context for living in a time of rapid and unremitting
change.

Why Now? Thriving in
an Unfinished Revolution

ASPECTS OF ARCHETYPAL TRUTHS that have been expressed in one culture can get lost for a time, then reappear in another where they seem like missing pieces of a puzzle that suddenly become available when they are needed. In times when people are struggling because their lives seem inextricably difficult, these verities can help individuals and entire societies regain the capacities they require to realize their potential. We are now in such a time, but to understand this, it is important to recognize that the archetypal truths discussed herein originally were filtered through the assumptions of ancient Greek mythology. To be useful to heroines today, details of such old myths must be updated so that they are relevant to the particular challenges we face.

The terms *myth* and *mythology* refer to expressions of archetypal stories in particular times and places.[1] Such stories provide us with a view into how archetypes were and are expressed in different cultures. The Eleusinian stories are part of the larger whole of Greek mythology, and like all mythologies, these may have been interpreted literally, especially by the uneducated (e.g., there really was a Demeter, and everything in her story happened, and that is when we began having seasons). However, the spiritual truths in wisdom traditions generally are revealed through parables, metaphors, or symbols, while political and economic truths usually are communicated through narratives that provide examples of broad principles or sets of ideas.

Our own times and places have their own mythologies that communicate cultural values while also reflecting universal truths. The story of the founding of the United States, for example, as well as accounts that reaffirm the American dream of opportunity for all, are aspects of the national mythology of my country. Although the United States now is a very religiously as well as ethnically diverse nation, its mythology and unconscious assumptions remain heavily influenced by Protestant Christian values. Thus, even people with secular beliefs typically assume, for example, the primacy of the work ethic.

In times of transition, culture wars often develop. In the United States today, our politics are mired down with conflicting mythologies that differ profoundly about whether or not recent cultural changes are threats to be curtailed or advances to be supported and encouraged. In a recent election, one group employed slogans about change we can believe in and the other, about taking back America. The continuing stalemate in our politics reflects a deep ideological divide over whether government and business should be all Zeus, with the emphasis on defense and economic competition, or Zeus balanced with Demeter's concern about caring for people. In the Middle East right now, competing mythologies are tied to different traditions within Islam and also to the tension between groups that want to join the rest of the world and those that wish to preserve or retreat to an ancient way of life.

The Eleusinian Mysteries were a change mythology of their time. Thus, we are confronted with one of those important cultural opportunities where the present has much to learn from what has been a missing legacy from an ancient past. The Eleusinian stories resonated with the citizens of Athens at a time when they were incubating a new future, just as our society is today; virtually everything seems to be in flux, which has been true for us since the 1960s. We even have our own version of a famine: climate change and other elements of environmental devastation.

By the 1970s, women were meeting in consciousness-raising groups where they talked about their lives and aspirations. From the gap that was apparent between what they were experiencing and what they dreamed of, a mass movement was born that was not dissimilar from the ancient Mysteries in its openness to different perceptions about what participation meant. This movement gained such powerful traction that it radically changed laws and social norms.

Recently, another new generation of women has again been recognizing a disconnect between their ambitions and their realities. As in the 1960s and 1970s, all this started with women talking, with the added benefit that in the digital age, these conversations would expand through social media. Soon women were blogging about the gap between theoretical equality and its realization. As a flurry of new articles and books have appeared, contemporary women's concerns are being discussed within the framework of larger current trends and issues. Women are not forming a special interest group in competition with other historically underrepresented groups or with men. Rather, they present women's issues as part of a larger push for a fairer and more just society for everyone—a partnership society, which is what the women of ancient Greece also wanted.

Although our lives are radically different than those of ancient Athenians, the Eleusinian Mysteries have much insight to offer us. Throughout the ages, women traditionally have been responsible for the maintenance of four major human activities—care, love, social life, and spiritual life (represented by the characters of Demeter, Persephone, and Dionysus in the Eleusinian narratives)—leaving men to focus primarily on success in the economic and public spheres, protection of the home, public safety and national security, and the creation of policies and laws to govern areas that women took care of in practice (represented by Zeus in these Mysteries).

The great challenge of our times is a side product of the opportunity women and men now have to escape rigid gender and other limiting role definitions. Most of us today want to express the fullness of who we are and what we can and wish to do. This reality was only dreamed of by those initiated in the Mysteries. Yet trying to enact this dream in real time in a society that is still designed as if women were home and men at work all day, when life no longer is that simple, can be stressful. It is especially difficult to manage all this when women still are expected to be "nice" and men to be "tough," in a world that requires that we have the flexibility to respond with what situations demand.

Moreover, while women have access to multiple roles, areas of life for which women traditionally have been responsible tend to be devalued in comparison with those historically assigned to men. As a result, both men and women today gain status by prioritizing competitiveness and accomplishment in the public sphere over caring, intimacy, spiritual depth, or enjoyment, resulting in a society and individual lives that are sorely out of balance. Yet these undervalued human activities remain important to individual and societal success and happiness. This is one reason that so many people today, especially women, find that their multiple roles require more than twenty-four hours a day to accomplish.

While I previously have written books on the hero's journey (*The Hero Within* and *Awakening the Heroes Within*), using the term *hero* in the generic sense to include both sexes, the strengths and values that typically have been nurtured by women are most critical if we are to achieve a balance of the masculine and feminine in our society. I use the term *heroine* to emphasize this, even though the developmental path described in this book is, at root, a human journey available to any and all of us.

While this remains an exciting time for women, given how much freedom and how many options we have (at least in most developed nations), implementing these choices in real time is demanding. It can be comforting to recognize that feeling overworked and overwhelmed is *unlikely* to be how things will be forever. Once the revolutionary changes in gender expectations that have shaken up our society have fully taken hold, new structures and norms will stabilize, and life eventually will be simpler. Imagine how much easier it would be if work and school hours were changed to conform to the needs of the new normal of dual career couples and many single parents. But for now, *Persephone Rising* can help you thrive within a revolution that remains unfinished.

The Eleusinian Mysteries appeared at a time when patriarchy was solidifying its grip on human consciousness. If we take an extremely long view, we can celebrate that the legacy the Eleusinian founders were working to hold on to has returned in upgraded form because of the efforts of many courageous women over time, and could be realized completely in the near future. The advantage of chaotic times in a digital age—where virtually everything has sped up, interdependence is real, and everything is in flux—is that in this chaos, change actually can, and does, happen quickly. Its direction may be difficult to predict, simply because every country, every business, and every person is influencing this change. That is why so often even major events come as surprises, such as when the Iron Curtain came down, apartheid ended, the 9/11 attacks happened, or the economy tanked in 2008.

Just holding the concept of an unfinished revolution can keep us grounded in reality while remaining optimistic. Many women today, even in high places, were raised by mothers who told them they could be anything they wanted to be. In truth, many of these mothers were saying this with their fingers crossed behind their backs, hoping or praying that life would be smooth sailing for their daughters. Of course, these messages helped their daughters be optimistic and confident enough to achieve great things, but the side effect was that when their boats hit the turbulent waves of backlash or other gender bias, those daughters often felt blindsided or even duped, which also might happen soon to their granddaughters. It is easier to sail through a storm when you anticipate that it might occur and are prepared.

This is a time that calls us to be heroines—sometimes just to get through the day, or to avoid the temptation to do things that are not right, or not right for us, in order to please others. As a person who wants to be happy, as I hope you want to be, I have noticed how many people think of heroism as inherently sacrificial, requiring men and

women to do something huge to save the world. The Eleusinian narratives come out of an alternative tradition—one that shows heroines
and heroes how they can make a difference without sacrificing their
own fulfillment and well-being.

It is true that some of us are called to very challenging heroic mis-
sions and find joy in them when they truly are ours to do. But it is
equally true that because we live in interconnection with one another,
doing our part in a way that is true to who we are, even if that part
looks small by society's standards, also is heroic. Such acts create ripple
effects that collectively make a huge difference, while also being so
right for us that we feel deeply fulfilled. One of my favorite characters in literature is Bonanza Jellybean from Tom Robbins's novel *Even
Cowgirls Get the Blues,* who asserts that heaven and hell are real, and
whether or not they exist in an afterlife, we experience them right
here on Earth. "Heaven is living your dreams," she says, and "hell is
living your fears."

When we grow into our full individual potential, our challenge
is to bridge the gap between our dreams and our current experience.
Yet our capacity to do this is intertwined with the consciousness and
potentials of our time. Collectively, we are continuing to act in ways
that result in getting what we do not want, like war, social injustice,
and environmental devastation, as well as in lives that feel harder than
they need to be. When enough attitudes and behaviors change, the
world we experience will change, too. Thus, the awakened heroine
becomes the best she can be, not so much to triumph over others, but
to contribute what only she can to realizing our human potential.

Doing this successfully requires expanding the more traditional
feminine virtues of Demeter to apply to the public as well as the
private world and opening to notice the deeper feminine wisdom
now arising from the unconscious and becoming available to our
conscious awareness. These together heal Zeus's arrogance, helping
him to enjoy collaborating rather than trying to control everyone
and everything. The fruit of this partnership between historically
masculine and feminine strengths gives us access to deeper, archaic
aspects of Dionysian joy.

As you begin to explore the stories of Demeter, Persephone, Zeus, and Dionysus in the following chapters, you will see how all four embody human qualities that are needed for wholeness today. Each narrative moves its characters from a relatively unconscious state to a more mindful expression of their essential gifts and offers challenges that spur them to grow and develop. The ultimate happy ending of their stories, however, comes from the interplay between them, just as today, in a radically interdependent world, our happy endings come not from simply getting what we want, but rather from the way our desires and those of others—and wild-card chance—come together.

The Conclusion offers you a toolkit to help you apply the narrative intelligence you have gained through reading this book. It will help you to protect yourself and others from internalizing limiting or harmful narratives, to craft narratives that highlight what is real and meaningful in life, and to reframe stories to transform attitudes and situations.

Part One

DEMETER

Demeter and
the Way of the Heart

FROM ANCIENT TIMES, men have, for the most part, possessed the physical advantage when it comes to strength and have excelled at hunting and war and other ways of solving problems with force. Women's strengths traditionally have flourished in the arenas of bonding, communicating, and understanding children, men, and other women, so that they could bring out the best in them. Even today, research on communication styles tells us that, overall, women's communication has a goal of connecting and relating in a more equal way, while men's is more competitive and motivated by a desire to impress.

I'm not implying here that either sex is better than the other. It is clear that the world needs both because the survival of the species depends on it, and in truth, men and women, when living in loving partnership, make one another and others happy. Men and women complement each other's strengths well. However, when a male perspective dominates female ones, the world ends up living narratives that may be successful in some situations but simply cannot get us

the results we want in others. For example, if we want peace, why
do we keep telling war stories? Why don't we turn to the half of the
human race that has fostered other means of resolving conflict? Force
can stop violent behaviors temporarily, but authentic sharing through
story, which often has been nurtured by women, can move antago-
nists toward understanding one another and building the trust that
leads to lasting peace. Similarly, in our politics, warlike competition
prevails when candidates run for office, but to govern successfully,
they need to utilize more feminine modes, reaching across the aisle to
solve problems together.

All of the major religions in the world instruct us to love one
another as a road to a better collective and personal quality of life.

- ✦ Jesus repeated this decree over and over, in slightly different
  words: "A new command I give you: Love one another. As I
  have loved you, so you must love one another" (John 13:34, NIV).
  "If you love me, feed my sheep" (adapted from John 21:17). And
  quoting the Torah, "Love thy neighbor as thyself" (Lev. 19:18;
  Matt. 22:39, ASV). It was his major message.

- ✦ Rabbi Sefer Baal Shem Tov, founder of Hasidic Judaism, spoke
  to the deep roots of love in the Hebrew faith: "'Thou shalt love
  thy neighbor as thyself.' Why? Because every human being has
  a root in the Unity, and to reject the minutest particle of the
  Unity is to reject it all."[1]

- ✦ The sayings of Muhammad, selected and translated by the Sufi
  Kabir Helminski, include the very strong statement, "You will
  not enter paradise until you believe, and you will not believe
  until you love one another."[2] Rumi, the thirteenth-century Sufi
  mystic and poet, proclaimed, "It is Love that holds everything
  together."[3]

- ✦ The Buddha enjoined us to "radiate boundless love towards the
  entire world—above, below, and across—unhindered, without
  ill will, without enmity."[4] Loving-kindness remains a cardinal
  practice of modern Buddhism.

✦ In the Hindu tradition, love also is the religion's central tenet. Swami Sivananda sums this up in these words: "Your duty is to treat everybody with love as a manifestation of the Lord."[5]

Of the many aspects of love, these spiritual injunctions promote personal kindness and altruism, both of which are part of Demeter's archetypal character. Yet learning to love one another remains an evolutionary challenge for us as a species and for each of us individually.

If so many religions tell us that learning to love is the fundamental task for humans, why is it so difficult to do, sometimes even for the very religious organizations that claim to embody that teaching—a number of which have been responsible for great atrocities, and in too many instances still are? As patriarchy became established in the Mediterranean, male values and traditional activities developed into societal norms; balancing virtues, like caring, came to seem much less important. By the time of the major Greek philosophers, rationality, logic, and objectivity were valued over emotions. In fact, the ability to repress one's emotions in order to think clearly grew to be a social ideal.

Mara Lynn Keller, a professor at the California Institute of Integral Studies, notes that the Eleusinian Mysteries were critical to fostering love, mutuality, and community in ancient Greece, especially at a time when it had developed a strong class and patriarchal structure. "The rites at Eleusis were considered essential to the survival of humans," she observes. She quotes the Greek historian Douris of Samos, who, writing about the Eleusinian Mysteries, maintained, " 'The life of the Greeks would be unlivable, if they were prevented from properly observing the most sacred Mysteries, which hold the human race together.' "[16] Cicero, who spoke of his initiation into the Eleusinian Mysteries as giving him reason to live in joy and die with hope, concluded that they were the most divine institution from Athens that contributed to human life.[17]

I believe that the Eleusinian Mysteries were so powerful because ancient Greece, like our world today, was out of balance, valuing attitudes and behaviors associated with men more than those associated with women. The Mysteries compensated for this by offering experiences that evoked empathy in the initiates for Demeter's fierce love and protectiveness of her daughter and for her daughter's potential misery at being married off to someone she did not even know. This tradition reinforced the wisdom of the heart to include empathizing with someone before simply using laws or logic to determine his or her fate. Demeter's story, which you already know, provides a narrative that can be seen as a stand-in for many analogous situations, encouraging people to stand up for whom and what they love and care about, even if it means taking on the power structure and challenging laws that treat people like objects or resources, as if they and their feelings do not matter. In recent centuries, with the triumph of rationalism, the power of the heart has been devalued compared with the power of the mind. This leads many people to disregard the wishes of their hearts almost entirely, making it difficult for them to find happiness.

Contemporary science is demonstrating that the emphasis in most religions (and in much of literature) on making choices with the heart is not just a fanciful metaphor. The heart is not just a muscle, as formerly thought. Modern neurocardiologists now view the heart as containing within it a little brain of about 40,000 neurons that conducts a flow of information throughout the body, which furthers emotional intelligence and empathy. The HeartMath Institute[8] has demonstrated that people who show loving and other positive emotions promote "positive coordination" between head and heart: just thinking of what you love and makes you happy influences your heart, which in turn influences your body. The result is greater happiness and health, with much less stress and fatigue.[9]

Many emotions are felt in the body, which is why we have expressions like warmhearted, softhearted, or hardhearted. Whether or not our hearts literally get warm or soft or hard, these metaphors correspond to actual feelings in the chest. Let's say you are watching a child that you love who is sleeping and looking very sweet. If you pay attention, you actually might sense warmth in your chest in the area of your heart. You also might feel your muscles and tendons softening. If you remain there for a while, perhaps you will notice that your whole body begins to relax. Not only that, you might find that just remembering your child sleeping can have the same effect. Conversely, you may experience a moment when someone does something that angers you and you feel a tensing up in your chest as you withdraw empathy from this person whom you now perceive as a threat. Very soon, your entire body might tighten up. Similarly, while our physical hearts might not get literally broken, a person is likely to feel physical pain in the area of the heart in times when they are deeply wounded by what someone has said or done or when experiencing grief. However, many of us repress the sensate feelings we experience in and around the heart because we have been told a story—that such expressions are just metaphors—and we believe the narrative rather than our felt experience.

Researchers generally used to try to filter out such visceral feelings that carry information about emotions, but today those in the social sciences also are recognizing that the pretense of being objective can simply blind them to their subjectivity. Increasingly, it is considered wise to get in touch with the values and feelings that you have about the subject, and especially what you want the outcome to be. The more conscious you are of your feelings, including those in your body, the more likely you will be to avoid unconsciously biasing your results. Many now are explicit about such matters in their research reports. Similarly, organizational development experts such as Daniel Goleman (author of *Emotional Intelligence: Why It Can Matter More Than IQ*) are urging business leaders to develop their emotional and social intelligence, which includes connecting the heart, mind, and body, so that they actually know what they are feeling and can read and appropriately react to the feelings of others.

Women's traditional roles have required us to develop emotional and social intelligence. For example, caring for young children before they can articulate their needs builds our capacity for empathic attunement to body cues, as does trying to figure out what men, socialized to be more stoic, need and want. And even when we move into less traditional roles, these abilities still are expected of us and useful to us, even if they often are undervalued.

Dimensions of the Demeter Archetype

Demeter is a goddess who embodies openheartedness, meaning someone with a welcoming, generous attitude toward other people. Demeter's story illustrates her journey to become equally caring and generous to herself, as well as others, and to balance an open heart with an open and thoughtful mind. The impulse to care for one another may explain why our species survived when Neanderthals and other hominids did not. Anthropologists have found fossil-bone remains that reveal that very early in human development, our ancestors took care of one another, even of those ill enough to need extended care. Such kindness is thought to have improved our survival odds (as fewer of us died and care cemented loyalty), just as learning to solve conflict peaceably and to nurture one another and the earth can do. In our own time, people are less likely to become homeless or sink into depression if they belong to a caring community. Equally important, loving and being loved make people happy.

Demeter personifies the life energy that makes things grow and that heals our bodies when they are hurt, as well as mother love and, by extension, any pure, unconditional love. She helps us remember our interdependence with the earth and one another. As the grain goddess, she often is depicted in ancient Greek art majestically holding up a sheaf of barley. At the pivotal moment at the close of the Eleusinian Mysteries initiation rite, a priestess, embodying Demeter's presence, would raise this evocative symbol. By the time she did so, the sheaf of barley had attained such significance—symbolizing as it did the essence of Demeter's teachings—that seeing it had a powerful impact on those assembled. Demeter's teachings emphasized loving and caring ways and a connection to the natural world and other people. As the goddess of agriculture, she was credited with teaching humans to grow their food, rather than just hunting and gathering, in order to create greater prosperity.

During a long period in history when men and women were perceived to be living in different spheres, she presided over the private world of family and close social ties and the care of children, the elderly, and others requiring assistance, all of which fell within the domain of women. Because of their willingness to sacrifice for the good of their offspring, mothers typically have served as models of selfless love. As earlier referenced, Demeter's name is derived from the Greek root word *mater*, which means "mother."

The fact that babies are born out of women's bodies and are dependent and generally adorable fosters an attachment bond that makes many mothers love their children as much as they love themselves (and often more). For some, the vulnerability of loving someone else so intensely builds empathy that then can result in their loving others as much as themselves. Demeter embodies the power of this love, not only in her feelings for Persephone, but also in the ultimate way she expands her love and care to humankind through the creation of the Mysteries.

In real life today, parents partner in raising children, sex roles are not always divided so clearly between men and women, and psychologists frequently emphasize how important a father's love can be to a child in fostering self-esteem and in building the confidence that both boys and girls need to be able to succeed in life. A caring father is important for boys as a model for being manly without being macho and stoic, and for girls so that they are less likely to fall for an abuser and more likely to seek out positive male mentors who can help them succeed in their careers.[10] The nurturing father that comes to mind when anyone uses the term *daddy* embodies Demeter strengths reflected in manly forms.[11]

All archetypes have their not-so-pleasant shadows. In the deep recesses of archetypal images of mothers is a mother goddess that is more primal than Demeter and that is associated, like the goddesses Kali or Isis, as much with the grim reaper as with the stork. Depth psychologists talk about the "devouring" mother, whose womb also is a tomb, because her nurturance can enable crippling dependency and enfeeble its recipient. A descendent of this archetype is the evil stepmother who would very much like to get rid of her stepchildren, perhaps by sending them off into the woods and hoping they get lost.

In modern life, the negative mother shows up in martyrs who use

guilt as a form of control; or who live through their children, overwhelming them with care and attention, thus smothering the children's authentic spirits; or who worry about their children so much
that they continually critique everything they do to try to fix them,
and in so doing erode their self-esteem. Demeter's shadow also is there
when men or women are so all-giving that they neglect to take care of
themselves and turn into martyrs, gradually becoming so empty that
they dry up and are consumed with bitterness.[12]

The Demeter Story: Then and Now

The opening story of this book *is* Demeter's story, so I will not repeat
it here. Stories in wisdom traditions have multiple layers of meaning.
Demeter's explained the origin of the seasons (likely to children) and
of the Eleusinian Mysteries, similar to the way we tell stories today
about how this nation or a company got started.

Reading the narrative, what immediately jumps out is that it is a
love story, but instead of it being about the love between a man and
woman, it is about the deep affection between mothers and daughters.
My personal link with Demeter, which I already shared, came from
the primal human reality that a life filled with love and caring opens
us up to the loss of whom and what we care about. At a metaphysical
level, the story can be interpreted as a mother's confrontation with her
daughter's death, with the daughter's return signaling resurrection,
reincarnation, or mother and daughter being reunited in an afterlife.
On a psychological level, the myth can be viewed in the context of
the great challenge that parents experience, having invested so much
in raising a child who then grows up and leaves them. In this meaning, Persephone's sharing her time between Hades and Demeter could
parallel a girl growing up and getting married or otherwise becoming
involved in her adult life but staying very connected to her mother.

This level of the story relates to a primal human reality, while
Demeter's and Zeus's standoff and its political significance for gender
roles in ancient Athens and today reflects continuing difficulties in
relative male/female power that nevertheless are culture specific and

fixable. The plotline of Demeter's primary narrative follows the basic pattern of a mystery story, with a twist. It begins with a mother whose adolescent daughter is missing, and Demeter is the sleuth seeking to solve the mystery of where she is and how to get her back. The plot begins with the discovery that the child is missing, so the first mystery to be solved is, Where is she? Then Demeter learns from Hekate and later Apollo that her daughter has been kidnapped by a man who is in cahoots with the child's father and now is being held captive somewhere under the earth. In the normal mystery-story form, she then would engage law enforcement to find the daughter and arrest the father and the kidnapper. But the father is the law! What now? The new mystery to be solved is how, in such circumstances, to get Kore back.

This plot shift is both radical and very modern. We live in a time when we know that some of our leaders are corrupt and cannot be trusted, and many of our heroes are those, like Demeter, who stand up to them. Beginning in the 1950s, we have seen emerge, one after another, liberation movements that address a dilemma similar to Demeter's. Their desired outcome has been not to overthrow the powers that be (as in violent revolution), but to bring them around. In this nonviolent tradition, movements are built by organizing public events (e.g., marches, sit-ins, or strikes) designed to raise awareness of injustice, which then engender sympathy from the larger public. Eventually, the growing consensus, within a democracy, is powerful enough to produce a change of heart, or at least of behavior, in those in power.

As with the tradition of nonviolent revolution, the radical twist in Demeter's story requires a change in attitude by the mortals, the gods, and Zeus. Demeter is not merely a lone mother trying to retrieve her daughter; she is challenging a father's right to decide unilaterally whom his daughter will marry. When Demeter remembers that she is a goddess, it occurs to her to use her real power. Everyone just needs to be reminded that she has it—hence the famine. There is no way for us to know this, but I would wager that over the years, more and more initiated mothers and daughters in ancient Greece used the story to pressure fathers into consulting them when arranging a marriage.[13] Male

initiates may have gotten the message as well just from paying attention to the implications of the story for how they should, and should
not, treat the women in their lives. If we remember that Demeter's
story fueled a movement (the Mysteries), we can see how her example
might have motivated individual rebellions—contemporary versions of
Rosa Parks refusing to give up her seat in the bus.

Ancient Greek theater had two major forms: tragedies, which end
tragically, and comedies, which end happily, whether or not they are
humorous. The classic happy ending is the restoration of community,
which occurs in Demeter's plotline when Persephone returns to her
and they both rejoin the community of Olympian gods. We see this
spirit today not only in the tradition of nonviolent revolution but also
in the growing use of restorative justice in our legal system.[14]

The following sections of this chapter examine the various episodes
in Demeter's plotline that lead up to the happy ending and that provide lessons for us in our own time. They illustrate how you can

+ be strengthened by a connected form of consciousness;

+ learn to cope with loss and fear and become more resilient;

+ resist the devaluation of Demeter qualities in yourself and the
  world;

+ leave circumstances where you or others are demeaned,
  mistreated, or undervalued; and

+ claim the power to stand up for yourself and your values.

The Demeter Archetype
Mythological Demeter: the mother goddess, Earth Mother, and grain goddess
Primary Heroine Lesson: the ability to live a connected life of mutuality and reciprocal care
Narrative Progression: moves from serving others to reciprocal care

Gifts: compassion, altruism, generosity, self-care, and care of others
Historical Gender Association: with the feminine in men and women
Decision-Making Mode: the heart with input by an aware mind
Inner Capacity Developed: an awakened, wise, and connected heart
Counterproductive Form: the bitter martyr who never got what he or she wanted

Demeter and the Eleusinian Promise
Primary Happiness Practice: showing kindness and investing in social relationships
Contribution to Prosperity: care of people and resources, partnering, and generosity
Contribution to Freedom: feeling one with the world, unafraid of life's cycles

APPLICATION QUESTIONS:

Based on what you know so far about Demeter,

Do you have too much, too little, or not enough of this archetype?

If it is present in your life, was it a family legacy (i.e., you were taught to be like that and are), a vocational ally, or something deeply and authentically you?

Is this archetype something you have been taught or learned you "should" or "should not" be?

Do you like people who reflect Demeter's stance in the world? Why or why not?

Demeter Lesson One:

Living a Life of
Connected Consciousness

D EMETER WISDOM VIEWS all of life as a whole, with each of us as a microcosm of the greater macrocosm. When women were gatherers and men hunters, the earth was seen as the mother and humans as her children. The lens of a gatherer is different than that of a hunter. The hunter needs focused attention to target possible prey and respond to danger. The gatherer surveys the territory for multiple treasures, which include food and other raw materials that can be used to create enjoyable lives, looking at the world contextually.

The earliest versions of the Eleusinian Mysteries were women's fertility cults that placed value on the cycles of natural and human life. The influence of such beliefs is manifest in the Demeter and Persephone story, with its emphasis on the creation of the seasons of the year, reassuring people that spring indeed will follow winter, and yes, the grain and other vegetation will sprout in time to avert famine. This tradition also venerated the reproductive cycle as similarly sacred—menstruation, sexuality, birth, nurturing growth, and death.

The connection with nature was felt viscerally in women's lives. Women learned to view their menstrual cycles coinciding with the phases of the moon as a sign of their oneness with the world. They linked the reproductive aspect of conception with the planting of seeds in the earth, and birth with the new plants sprouting, and they valued the male role in sex and in protecting and caring for a pregnant mate and children. Death and burial also were connected metaphorically to planting in a context that assumed that just as seeds grow into new forms, people too would grow back in new bodies—on the earth or elsewhere—providing hope that death was not the end.

In Demeter, we find a goddess who combines a complex of qualities: caring for the earth, overseeing and teaching about agriculture, and showing love as care and compassion for others and oneself. The essence of Demeter's consciousness is the ability to be fully centered in herself—knowing who she is and what she wants—and connected to others and the larger world, which requires giving and receiving as a normal part of living. When you achieve this, you may experience the end of alienation and begin to feel a radical sense of belonging and connectedness.

Demeter consciousness also is affected by women's historically, and even now, having the primary responsibility for caring for infants and young children. You can think of women, throughout time and today, trying simultaneously to get work done and be aware of what their children are up to, being sure that they are not in danger or otherwise doing something they should not be doing. This requires an ability to multitask, which may be why, on average, women have more connective tissues between their brain hemispheres than men do.

Being able to track a variety of things going on around us fosters the development of the kind of cognitive complexity that researcher Robert Kegan (*In Over Our Heads: The Mental Demands of Modern Life*) says is needed to cope with the complexity of modern life—individual, corporate, or intergovernmental. Educators used to define some children as field independent, meaning they can shut out what is going on around them, or field dependent, meaning they cannot. When they recognized that boys often were in the first category and girls in the second, they reframed the term "field dependent," with its assumption of dependency, and began to use the more accurate term "field sensitive." Many women with an awakened Demeter archetype also have strongly developed empathic abilities, so they are feeling, not just noticing, what is happening around them, and that feeling is what motivates the helping response.

When our hearts are open, it seems natural to want to thrive together. We view our needs and desires in a wider context of the needs and desires of others. This shifts the focus of our brains and allows us to see the interlocking needs of people in our families, communities, country, and, increasingly, the planet. When we look at nature, we observe interdependent ecosystems, and from that, we gain the ability to understand the economic and social worlds as interdependent human systems, where that which affects our personal good also is happening with and to others. We see this capacity to feel and think in cognitively complex ways in the current focus on the environment, the discussions of women's issues, questions regarding the effectiveness of our schools in educating the young, political debates about our responsibility to care for ourselves versus caring for others, and the reemergence of spirituality in religions and individual human lives.

The examples in this lesson that follow show how Demeter's connected worldview remains with us—in human behavior, literature, moral development theory, and contemporary leadership styles—and provides models for experiencing this in your own life.

Archetypes often form the basis for enduring literary works, such as Virginia Woolf's novel *To the Lighthouse*. The central character, Mrs. Ramsay, is a beautifully portrayed Demeter-like character, viewed from the perspective of an early-twentieth-century writer, that many believe was modeled on Woolf's mother. Mrs. Ramsay is a mother of eight children, married to an academic philosopher who is more concerned with his reputation in the field than with his family. The secondary character, Lily Briscoe, is an artist who is painting a portrait of Mrs. Ramsay with her youngest son, James. Lily sees Mrs. Ramsay as an archetypal figure, and paints her subjects as a kind of modern Madonna and child.

Woolf's literary portrait is her tribute to qualities that women in traditional roles have continued to embody. Lily's goal is to escape those roles to focus on her art, but she, nevertheless, still values them. Lily greatly admires Mrs. Ramsay and worries about her. As she sees it, the male figures in the book are tied up in their egos and come to Mrs. Ramsay as if she were a fountain from which to drink, in order to fill themselves up with her life-force energy. The reader perceives what Lily is concerned about, but also that when Mrs. Ramsay's energy ebbs, she knows how to recharge it—and this is where we can learn from her. She moves her attention outward and is restored by taking in the beauty of nature or feeling her love for others. In one scene, she is depleted and sad, thinking about the meaninglessness of life, but then turns her full attention to the light over the water from the lighthouse. Suddenly, she feels such oneness with it that she experiences ecstatic joy. This is not the escape of transcending oneself, however. Rather, Mrs. Ramsay finds connection with a larger sense of oneness that emanates from her heart, an experience that centers her in herself. Indeed, what Woolf depicts as a feminine gift can act as a powerful antidote to feelings of alienation in both men and women—and potentially in you and me as well.

Mrs. Ramsay's particular magic does not emanate just from this ability to reenergize through connecting with the beauty in the world but also from her capacity to help both men and women escape from their egotistical insecurities to truly connect with one another and, in that way, to catch some of her connected consciousness. In a famous dinner scene, Woolf takes the reader inside the mind of each of the guests. Each one is cut off and alienated, many of them fearing that their efforts will not be received and valued, especially Lily with her art and Mr. Ramsay with his fading scholarly reputation. Mrs. Ramsay appears oblivious to the discomfort around her, chatting on about what seem like the charming trivialities of a woman of her day. Yet she knows the stage has been set: the food is delicious, and her conversational style is lightening them up and inviting each into the conversation. She and others have been worried about a young couple not yet returned from an outing, but then they appear, clearly having declared their love for one another.

Their appearance creates a tableau, or image of love, that has the potential to help everyone let go of their self-involvement and form a community. As the sun is setting, Mrs. Ramsay knows the time is right and tells the children to light the candles. From that ritual act, suddenly a "change at once went through them all . . . and they were all conscious of making a party together in a hollow, on an island."[1] At this moment, Mr. Carmichael, a particularly curmudgeonly guest, turns to Mrs. Ramsay and bows, clearly recognizing the magic in what she has accomplished.

What a wonderful gift to have, whether we are hosting a dinner party, leading a work team, teaching a class or workshop, helping our families to get along, or getting members of Congress to work together![2] A ritual act of beauty can transform the energy around the table, and in this way shift the inner stories people are experiencing from insecure "it is all about me" narratives to "I am part of an us" ones.

Woolf's beautiful example shows us how Demeter's quality of connected consciousness provides an antidote to alienation, depression, and anxiety so that we can feel at home in the world. Experiencing that level of comfort with others and in the world requires finding the match between that place in you where "your deep gladness and the world's deep hunger meet," as Frederick Buechner so eloquently puts it in *Beyond Words: Daily Readings in the ABC's of Faith*.[3] To achieve this, we need to develop enough of Demeter's empathy and altruism to discern what people might want and be motivated to give it to them. This centered but connected state of thriving that Buechner evokes is neither selfish nor selfless. Selfish people rarely thrive, for they miss out on the sense of meaning and purpose and the wonderful kind of happiness that comes when we know that our lives are making a difference to others. Selfless people can be like the proverbial codependent—the drowning person who sees someone else's life flashing before her eyes.

Discovering your optimal match with the needs of the world is very different from twisting yourself around to be something you are not to try to fit a cultural ideal or preexisting role, like Cinderella's stepsisters struggling to cram their toes into the glass slipper. Instead, the optimum is to find a career path that fits us, like Cinderella's feet slipping effortlessly into the shoe held in the handsome prince's hand.

Gaining a sense of purpose requires a journey of discovery with moments when you realize what and whom you care about. Sometimes it is helpful to think of yourself as a puzzle piece looking for its place in the family of things, and then to experience joy at being just where you should be—not searching, not yearning, but in your right place. When you act the most authentically, you may fit into the space that is there for you, where your strengths and interests fulfill a preexisting need. Perhaps you have felt the immense satisfaction that comes in these very exquisite moments, or even longer periods of life, when you feel perfectly yourself—the right person, in the right place, at the right time to do something that helps someone or makes a difference to a group or the world.

Such a consciousness moves beyond an us-them way of thinking, leading to an approach to morality that is very different from traditional rule-based systems and also different from more modern, postconventional ethical approaches where us-them polarities remain. Scholar and professor Carol Gilligan began research in the 1970s after her female students told her that while they understood Lawrence Kohlberg's widely respected theory of moral reasoning (*The Philosophy of Moral Development*), and could pass a test on it, it did not represent how they worked through moral issues in real life. Gilligan's now classic book *In a Different Voice: Psychological Theory and Women's Development* plus her subsequent work have had a major impact on the moral-development field, flipping many of its assumptions on their heads.

Kohlberg famously ascertained levels of moral development in his subjects by interviewing them about the "Heinz dilemma," a scenario in which a man named Heinz steals a drug to save his critically ill wife because he cannot afford to buy it. Kohlberg then asked those he interviewed, should Heinz have stolen the medicine, and why? Kohlberg's research, done primarily on Harvard students, posited that those with the most developed levels of moral sophistication would justify their position based on whether they believed human life was more important than property rights, or the converse.

Gilligan discovered that for the women she studied, moral dilemmas were imagined not so much as competing principles but as values existing in networks of relationships and in the complexity of the real world. Thus, they would suggest that reestablishing community through communication—such as seeking the aid of others who might loan or donate funds—could result in Heinz getting the help he needed to save his wife without breaking the law. Indeed, the women worried, if he stole the drug and was sent to prison, who would get the medicine for his wife if she needed more?

In this process, the women essentially critiqued the premise of the Heinz dilemma by (1) shifting the question away from stealing as the only possible intervention, thus expanding the context in which the issue could be considered; (2) assuming that Heinz should save his wife rather than leaving that in question; and (3) brainstorming things he could have done other than choosing between the options given them. Although I don't believe Gilligan mentions this, I notice that the women were utilizing narrative reasoning instead of dualistic logic: if he does this, what will happen, and then what? These questions set the framework that generated alternative stories.

Gilligan was one of the many women researchers who questioned an unconscious male bias in research. In many cases, research was done on men and then just assumed to apply to women. Kohlberg's team did interview women as well as men, but in deciding what the levels of moral development were, the results naturally reflected Kohlberg's own beliefs and values. The saddest part of Gilligan's findings came from transcripts of the original Kohlberg interviews, which consistently rated women's moral reasoning lower than that of the men, and even assumed that the women's answers showed a failure to understand the question. Gilligan reviewed the interviews and noted that women subjects would begin to falter and sound insecure as they felt misunderstood and sensed that their thoughts were being devalued.

Gilligan found, through reviewing Kohlberg's studies and in her own research, that a majority of, but not all, men tended toward the ethic of justice described by Kohlberg as the basis of their moral reasoning, while most, though not all, women described being drawn more to an ethic of care. Yet once Gilligan articulated how this mode of moral reasoning worked, many men said that this also was how they resolved moral dilemmas.

Gilligan assumed three major stages of moral development, as did Kohlberg, but what she discovered to be the stages of development within the ethic of care were different from those of the ethic of justice. At a preconventional level, women and men are most concerned with themselves and their survival. At a conventional level, they believe that goodness is self-sacrifice, as women have been socialized to believe for eons, so the value is doing what others or the larger society need. At a postconventional level, moral issues are resolved by finding solutions that balance care for oneself and care for others, with an underlying commitment to the principle of nonviolence—that one should not harm oneself or others.[4]

For our purposes, we can think of the ethic of care as the realm of the Demeter archetype, and her story dramatizes her development from the conventional to the postconventional feminine ethic, a transition from viewing morality as meeting others' needs at our own expense to recognizing that our needs count equally. Demeter moves from a self-sacrificing assumption that she has to keep making the crops grow, even when Hades has abducted Kore, who is the person she loves most, to recognizing that her own needs matter too, so staging the first ever recorded sit-down strike.

Gilligan offers the ethic of care as complementary to, not in competition with, Kohlberg's model. It has interested me to notice that, even today, many scholars distort Gilligan's work, viewing it through their own inner us-them filter. They then take Kohlberg's and Gilligan's models to be competing stories, arguing that women's ways of doing things are better than men's ways or the converse. Too often in the academic world today, scholars make their reputations by attacking other theories and then claiming the superiority of their views, with the result that academe becomes a battleground of competing ideas. Gilligan's work also reminds us that we still can have intellectual communities where human knowledge evolves as we share ideas and learn from one another. This is made easier when we recognize that the ethic of justice reflects a Zeus archetype view of the world and the ethic of care reflects Demeter's—and that we need both.

I find that in my personal life, I solve moral dilemmas with the ethic of care, but when thinking of larger social issues, it seems natural to employ both models. The former helps me understand multiple perspectives, and the latter helps me, when down to the wire, make tough decisions about my priorities. That being said, the unconscious cultural assumption that men, generally of an elite group, are the norms for thinking and doing against which women (and other men) should be measured has an emotional violence to it, like a sword that undercuts the confidence of those who do not fit the mold.

The standard approach to career development is to decide on a career path that is right for us, develop a plan for achieving it, and go for it with determination and persistence, an approach that can ignore the context of the needs of others in our personal web of connection. In *Composing a Life*, Mary Catherine Bateson argued that this more me-centered approach was traditionally male, especially in times when men just assumed that their wives would, and should, adjust to support the breadwinner's career (or best-paying work options). Bateson called attention to a complementary female career path that grew out of these assumptions. Women, by necessity, have had to look at wider family needs and then figure out how to pursue their careers based on the real-world constraints of spouses' jobs and the needs of children and other family members who require care. The image of "composing," as in music, highlights how Bateson saw this mode of career development as an art form requiring great creativity.

As women's incomes from working have become more critical to family finances (or in circumstances where their paychecks always were) and as women have become more career oriented, we expect our work to count as much as our partners' in career planning. Today, it is more and more normal for couples to consider how each person in a family can realize his or her potential in the context of the good of the entire family, and how everyone can fulfill their responsibilities as students, workers, and citizens to the larger entities that rely on them. I often hear women use music metaphors to describe their efforts to achieve work-life harmony (rather than work-life balance, which seems impossible to attain) and to orchestrate events so that others in their family can find such harmony too.

A much-discussed problem today for women leaders motivated by Demeter values is that their work frequently is evaluated by the most Zeus-like of men. Like Zeus, those men assume the rightness of their way of thinking and judge women based on it, not recognizing the value of a complementary history and view to theirs. In an early groundbreaking study of successful women leaders, *The Female Advantage: Women's Ways of Leadership*, highly influential writer and consultant Sally Helgesen identified a particularly female way of leading that I would regard as Demeter's: from the center, by inspiring, not compelling; by forming a matrix supporting a central mission; and by complementing organizational hierarchies with mutuality and collaboration.

If you are a person who reads management literature, you immediately might think, "Wait a minute, that is what male, as well as female, management gurus are now advocating because we are living in a globally interdependent world." Helgesen's more recent case studies of successful female leadership-style practices—at Intel; Beth Israel Deaconess Medical Center in Boston; *The Miami Herald;* Nickelodeon, the children's television network; and Anixter, a telecom infrastructure supplier—concluded that such companies represent the wave of the future (see Helgesen's *The Web of Inclusion: A New Architecture for Building Great Organizations*). Extrapolating from how these successful contemporary companies are run, Helgesen predicts that

> great companies will all operate as webs of inclusion in the future. Webs allow organizations to draw on the widest possible base of talent, a huge advantage in an economy based on knowledge. They allow resources to flow to where they're needed. They undermine the tendency to become hierarchical. They put organizations more directly in touch with those they serve, and make partnerships easier to achieve. Perhaps most importantly, they break down the old industrial-era division between the heads of organizations and the hands—those who come up with ideas and those who execute them. In doing so, they return joy, creativity and a firm sense of participation to the work done at every level.[5]

Leadership for Transformational Change, a project I co-led with scholar and consultant Judy Brown, in partnership with the Fetzer Institute, the University of Maryland School of Public Policy, and the International Leadership Association, came to similar conclusions about what is working today that results in positive transformative change. The project convened leaders and leadership scholars and educators over a period of three years with the task of determining what leadership approaches are most successful in the current environment. As reported in *The Transforming Leader: New Approaches to Leadership for the Twenty-First Century*, we concluded that effective leadership is collaborative and able to harvest fully the intelligence of everyone involved, but we failed to notice that, at least for now, many women are drawn more naturally than their male counterparts to the connected quality of consciousness that informs such contemporary leadership styles.[6] Of course, women have this advantage only to the degree that their way of operating is partly informed by the Demeter archetype. This understanding clarifies how it is that so many men can lead in this way too, motivated by their own inner Demeter qualities.

That being acknowledged, a recent review by Alice H. Eagly and Linda L. Carli of a research index compiled by InPower Women substantiates the continuing existence of a woman's leadership advantage: "Women's styles of leadership match up more with sought after contemporary views of leadership such as strong and collaborative relationships, teamwork, and ability to empower and engage workers. The study shows that female leadership is typically more communal and involves more communication. Leadership styles were found to be particularly gender-specific. Women were more interpersonal and democratic whereas men were task-oriented and autocratic."[7]

However, this advantage can render those of us with an awakened Demeter archetype prone to heartbreak. Women and men who have open hearts feel things deeply, so they can be even more devastated by the loss of someone they love than would those who are more stoic (like Zeus) or more able to quickly let feelings go and move on (like Persephone or Dionysus). Moreover, if we, like Demeter, live in a web of interconnection, we may feel the pain of things that are happening to others we do not even know and have a propensity to tear up when watching the nightly news or reading the newspaper.

The next section shows how a connected consciousness, combined with an open heart, also can bolster our capacity for resilience, giving us the courage to face whatever happens. Before going on, you may want to shore up your own reserves by experimenting with meditating in a Demeter-like way.

AWARENESS PRACTICE:
Experiencing a Connected Heart

Sit in a comfortable position, breathing deeply, and imagine your breath going down from your feet and bottom (like roots) to connect lovingly with the earth, appreciating all it brings us. Imagine that you can breathe up the earth's energy, which fills your body and goes out of your head like branches into the sky, as you lovingly experience gratitude for the beauty and wisdom of the earth and the heavens. Then, let all the energy of the sky, like sunlight, also fill your body, mixing with the energy from the earth. When this process seems to be complete, center in your heart, putting your hands over it, breathing into it, feeling appreciation and love for yourself.

Then, when you are ready, slowly envision your energy connecting outward, first to those you love, then to all those whom you know, and then gradually to encompass the world. Send your love out to the world, and then open your heart to receiving, allowing energy to flow back into you. As you do this, you can imagine love coming into you as you breathe in, and going out to the world as you breathe out.

Finally, give any energy that is more than you need to the earth by placing your hands on the floor or ground, or as near to it as is comfortable for you.

Demeter Lesson Two:

Demonstrating
Brave-Hearted Resilience

T HIS SECTION EXPLORES Demeter archetype wisdom about
how a loving and connected consciousness fosters resilience,
even in very difficult times such as a major loss. In the process,
it examines the tendency to worry the Demeter archetype gives us,
and shows how using story vigilance can help you shift from an anxious to a hopeful and confident inner narrative.

Having an open heart along with a connected consciousness makes
Demeter realistic about the dangers besetting those she loves. From
this, she inevitably becomes saddled with a predilection to worry.
Although such fretting can help her anticipate and prevent loss and
more everyday problems, some always will remain out of her control.
I've already shared with you how devastating it was for me when Doug
died but I felt less alone when I read about how, when she discovers
her daughter is missing, Demeter is overcome with worry and grief,
as any loving mother would be, and begins searching for her. She is
too distraught even to eat or bathe. It appears that some of the gods
and goddesses know what happened, but none will tell her, until the

tenth day of her search, when the goddess Hekate takes pity on her and reveals that Hades abducted Kore, and she is now in the land of the dead, a place Demeter cannot go. It was less than a day that David and I waited to learn what had happened to Doug, but those hours felt much longer.

Greek myths, as do most stories in wisdom traditions, tell us what happened but do not delve into the inner lives of their characters to let us know what they were thinking. We do not actually know what inner resources kept Demeter going during her difficult search, but we can extrapolate to what is known about what sustains people experiencing loss today. As mentioned in the Introduction, I drew strength from the modeling of a Demeter-like mother who remained open-hearted in the face of many losses. In addition, I was comforted by a dream in which Doug appeared to me surrounded by glowing light, telling me to make sure his dad gets the word that he still exists and is fine.

A dream such as I had or beliefs about being reunited with a loved one after death—in heaven, being reincarnated together, or, as from near-death testimonies, going through a tunnel and seeing light and those you most love who have died before you—provide solace that can allow us to move on. Sometimes consolation comes less from a belief than from a feeling of remaining connected with those who have departed. Some people, for example, continue to talk to them and occasionally even feel as though they are receiving an answer. For others, comfort comes simply from experiencing ongoing gratitude for the time they had with a loved one on this earth. David and I made a commitment to honor Doug by making the zest for life he embodied a part of how we lived, always remembering that Doug's favorite word, expressed with passion, was "More!" Somehow knowing that something of him lives on in us has been consoling.

In retrospect, I also realized that the need I felt to multitask to keep life together after this tragedy actually helped me. Drawing on findings from neuroscience, Ginette Paris, a noted mythology scholar and a colleague of mine, describes (in *Heartbreak: New Approaches to Healing; Recovering from Lost Love and Mourning*) how brain mapping demonstrates that love is necessary for babies and children to develop healthy emotional and cognitive capacities. The loss of love, even as an adult, impairs emotional and cognitive functioning until the brain rewires itself. Referencing triune brain (reptilian, mammalian, human) theory, Paris further explains that the loss of someone you love—through death or their just leaving you—can cause the mammalian brain to experience desperate longing and a willingness to do anything, even beg or placate, to try to get the loved one back. If the love of your life leaves you, you might beg them; if someone dies, you might beg God to make this be just an illusion, not real.

Meanwhile, Paris continues, the reptilian brain is furious and wants to lash out, which in Demeter's case is heightened by Zeus's casual and insulting lack of concern for Persephone's and Demeter's desires and feelings. This causes tension within the brain that is very painful and presses to be resolved. Paris argues that the tension between these two parts of the brain can be resolved only by the cerebral cortex, the distinctly human part of the brain, and by a massive brain restructuring, which requires consciously seeking new experiences.

If you ever have experienced a great loss, you may remember how difficult it was at first just to think about what to do tomorrow, how alone you felt, and how important it was to you when someone—or ideally many people—was there to support you. What you may not have noticed is how making burial and funeral arrangements, letting authorities and loved ones know what has happened, and then going back to work, paying bills, and getting reengaged in a social life began to restore your own emotional and cognitive capacities. As Paris advises, a big secret is to keep moving rather than give in to the temptation to fall into despair. This could be what happens when Demeter begins to wander in search of Kore. She undergoes new experiences that switch on new facets in her brain.

The biggest impediment to remaining open to life is the fear that if we do so, and if we then lose what we love, we will not be able to survive a level of suffering that we fear could be unendurable. But holding back from loving does not keep loss from occurring. The fear of suffering, however, can keep you from being fully alive and present, and from expressing your love. Knowing that you can survive the unthinkable can lead you to spend less time worrying and more time enjoying what you have. Yet, it is natural for mothers to worry about their children, however resilient they have become.

During the period when the Eleusinian Mysteries were at their height, fathers typically arranged for their daughters to be married at puberty. The grooms usually were men in their late twenties or early thirties who had had time to establish themselves financially before starting a family. Of course, those fourteen- or fifteen-year-old girls, like girls now, were, emotionally, still children. You can imagine how many of their mothers felt, with their little girls suddenly being taken away from them, living in another household, having sex with someone much older, and feeling overwhelmed with the abrupt change in their lives. If a marriage took a child far away, her family stopped being able to see her except rarely—and remember, there were no cars or trains or planes, no telephones, no cell phones, no instant messaging—and they had no immediate way of knowing how their daughter was doing.

Even now, it is such a powerless feeling, being a parent. You cannot be there all the time, and as you drop your children at the door of even the best day-care center, preschool, or elementary school—especially if they cry when you leave—there is always that nagging feeling: "Will they be all right?" Boys and girls are prone to recklessness as adolescents, since their sense of mortality has not kicked in. It can be frightening when they get a driver's license, or become sexually active, or go off to college, a job, or especially the military, or make unwise choices in adult life. And then there is the empty-nest syndrome, which both fathers and mothers can experience. We put all this love and effort into raising our children, and then it is their task to differentiate from us and live their own lives. That leaves an empty space that needs to be filled.

Even worse is the pain when adult children are ill, are mentally or physically disabled, can't find a job, or make a major mistake, become incarcerated, and then are unable to get work or housing because they have a criminal record. Such situations force us to confront questions of how much care and oversight we should provide. Ongoing worry is an impediment to happiness and can motivate meddling. It also can lead us to ask, "What did I do wrong?" If we see something worrisome that our child is doing or being affected by, better to ask what we can do to help, and whether we should help, than ruminate on our inadequacies.

Demeter's loss also can be interpreted to refer to your *metaphorical* children—art, products, services, organizations, and innovations of all kinds that you take pride in as you may in your children, if you have any. If your creations are devalued or the winds of fortune eventually destroy them or render them outmoded, and hence irrelevant, you may feel real grief. Those of us who write books eventually must put them out into the world, which might let them sink quickly into obscurity, get scathing reviews, or even be the talk of the town for a short while before they are forgotten. Even if they eventually are venerated as classics, we likely will not be there to see it. In the nonprofit sector, people put everything they have into developing a thriving institution, but a new board of directors then can decide that new blood is needed and let them go. In the for-profit sector, a company's founder may make out better financially but still need to move on when the board of directors says it is time to pass the reins. Or else the founder may find at some point that the firm's products or services are outmoded, and the business tanks. Even the president of the United States has at most only eight years before the world moves on. Life requires letting go, and letting go can be hard.

Men and women with a Demeter nature also may have parental feelings in roles such as teachers and managers that entail responsibility for others. When they cannot protect those in their charge, they may feel deep sadness as well as a sense of failure. My dear Aunt Harriet, a missionary who had no children of her own, experienced grief when she became so ill that she had to be torn away from the African children she was teaching and return to the United States. I've known managers who suffered strong feelings of powerlessness and sorrow when they could not protect those they supervised from mandated layoffs, plant shutdowns, or inhumane policies that they could not change but had to implement. Some people experience heartbreak when they lose the ability to work in the area of their calling, or when a business they started fails (as was the case with my dad), or have to "let go" employees who need their jobs in order to live. Yet it is the personal pain we endure that opens our hearts to those who have lived through similar heartbreak or worry and allows us to be there for them when they need it.

People with a connected consciousness generally are tracking all the possible dangers out there to those they love, which is why they are so prone to worry. It also means that relatively early in life they accept that death is real. What is best about such folks is that generally it leads them to focus on how to make the most difference to others and the greater world while they are here. This gives them the resilience to keep on trucking, whatever losses and disappointments they experience, to see how else they might make some difference. When she was elderly and living in a nursing home, my Aunt Harriet found meaning from figuring out about ways she could comfort the other residents and be helpful to the staff.

The heart is not only the seat of love; it also is where real courage resides. Your love may be for a person, a pet, a place, a landscape, a home, a mission, an organization, a community, or a vocational calling. When you face the vulnerability of knowing that any of these can be taken from you in an instant, you gain clarity about what you care about and what matters in your life. That is why it is helpful to any of us to imagine ourselves on our deathbed, looking back to determine what really matters. Once we have that figured out, many decisions can be made much more quickly, and the more likely it is that we will make a priority of taking time to be with and care for whom and what we love. If we do this, we also can spend at least as much time with our thinking defined by gratitude stories as by our worry ones. Such clear priorities also make it less likely that we will experience regret or remorse if we suffer loss for not having appreciated what we had when we had it.

Inside each of us, so say depth psychologists, is a young child (like

Kore, the maiden), an adult (like Demeter), and a wise older woman (like Hekate). A part of us may feel as vulnerable as we felt when we were young; it also is the part that holds on to children too long. If one's own inner child feels orphaned and bereft, that feeling is likely to be projected onto others, who will be perceived as fragile and needing help, allowing the helper to be reinforced in feeling strong. Our Demeter sides can become more resilient as we strengthen and heal the inner child, which we do by recognizing how vulnerable we actually feel and getting help when we need it. We also can trust the strength and wisdom of others more, so that we do not believe that we always have to help them.

Beasts of the Southern Wild, a 2012 award-winning film, dramatizes how Demeter wisdom can help even a child become resilient, an example that also can strengthen the inner child in us, as adults. Its heroine is a six-year-old named Hushpuppy, who lives on a narrow strip of land surrounded on both sides by bayous, in an area not far from New Orleans affectionately called the Bathtub by its residents. As in Venice, they are surrounded by water, which serves as their roadways, though unlike in Venice, life in the Bathtub is very basic.

This world mirrors an ancient, nature-loving community, reminiscent of indigenous life in times of plenty but poor by modern standards. The society seems in some ways idyllic: the different races live together in harmony, and the inhabitants celebrate with frequent festivals, trusting nature to meet their basic needs. They catch abundant fish, garden, hunt a bit, and live simply. In the first scene, they are holding a grand party, with fireworks and endless seafood they have harvested from the sea and cooked. Then we hear the voice of Hushpuppy telling us how much greater life is in the Bathtub than on the other side of the levy, where they celebrate rarely, while her people do so all the time.

A sudden, severe storm, a bit like Hurricane Katrina, floods the landscape, after which the waters recede a bit. The inhabitants who have not evacuated are relieved for a while, until they realize that the salt water has killed the vegetation, as well as animal life, making the area unlivable. At the same time, Hushpuppy, whose mother left just after she was born, recognizes that her father, Wink, is dying and that soon she will be an orphan. Hushpuppy tells herself a frightening story that she will be alone and have to forage in the woods, eat grass, and, in her words, "steal underpants." As the narrative unfolds, we see how Hushpuppy comes to shift her fear-based story to one that is more empowering, but first she encounters an additional terrifying tale.

Her schoolteacher has explained to her students that the polar ice caps are melting, which provides a larger meaning to the flood, connecting it for the film viewer with climate change and resulting severe weather. Depth psychologist James Hillman and Michael Ventura (in *We've Had a Hundred Years of Psychotherapy—And the World's Getting Worse*) declare that a major reason people are depressed is that we should be. We see signs of environmental destruction everywhere and homeless people in the streets. If we go to a psychotherapist, most of us will focus just on our personal issues and the impact of our families of origin. This by itself will not help, the authors emphasize. Even if our conscious minds are busy focusing on getting more of what we want, our unconscious selves are weeping at the sad state of the planet. We will feel better only when we get out of denial and address the real problems we see all around us and tell ourselves a can-do story of what part in remedying all this we can and will play.

Hushpuppy's teacher also told the kids a terrifying tale about ancient, monstrous frozen beasts that will be unleashed when the ice melts. To stress her point even more, she says that these beasts eat little children for breakfast and have no mercy. That is why the children have to learn to be strong. It is clear that she is worried about the children's future and wanting to fortify them with courage. Wink, knowing his time is limited, is panicked about leaving Hushpuppy behind. Like the teacher, he focuses on helping to build her strength, working hard to teach her macho ways of coping, all relevant to the Bathtub world that is dying—like fishing with her bare hands. But what he gives her that matters most is a sense of power to live a heroine's story, even having her repeat, "I'm the man," while showing off her muscles. Both teacher and father are providing important Zeus-like masculine support for Hushpuppy's next steps. However, he does not do what Demeter would do—line up anyone to care for her when he is gone. After all, she is only six, not sixteen.

Toward the end of the film, Hushpuppy realizes that Wink's death is imminent. He is in the process of dying, not eventually but very soon. And this is happening as everyone is recognizing that the Bathtub is becoming a wasteland, so the hoped-for return to normalcy is impossible. Hushpuppy is losing everything and does not know how to face it. In addition, she has been taught, and believes, that it is her responsibility to care for what created her, which is her father. Yet her first reaction is to run away from all this pain, and the journey she takes in the course of a day connects her symbolically and emotionally with much-needed feminine energy.

First, Hushpuppy and her friends go into the ocean, the very ocean that first gave them the sustenance they needed to live and then destroyed their home. Now it provides some healing. Ritually, this is going back to the original womb of the mother of all life, which also is one of the first steps in the ancient Eleusinian Greater Mysteries rite. She and her friends hitchhike onto a boat, hoping to find her mother. The pilot lets them out at a brothel, where a woman is kind to Hushpuppy, cooks for her, and dances with her, holding her close in a warm, motherly way. It is even possible this woman is her mother, but she does not take Hushpuppy in. As she is being held, Hushpuppy remembers how rarely she has experienced such comfort, but this little shot of Demeter energy restores her enough to do what she believes she needs to accomplish.

Second, although the film never tells us where she got her philosophy, Hushpuppy is strengthened during this time by a Demeter-like connected narrative and way of seeing the world, as well as a sense that, no matter how insignificant her life may seem, it and she both matter. Because such an implicit worldview comes naturally to so many women, it is possible that the new story even arises out of Hushpuppy's own unconscious mind. In the film, her voice-over shares her thoughts with us, and we learn that she believes she is a very small speck in the universe, but this realization is not dispiriting.

In fact, she feels awe at the universe's expanse and through experiencing a mystic connection with the whole of it. She further believes that in the interconnected universe, she can disrupt the whole by the ripple effects of something she has done, and her responsibility, then, is to make things right.

This take on personal responsibility is reminiscent of a Jewish idea that the world was created in a broken state, or shattered shortly thereafter. God then created people to perfect the world through acts of kindness, which can restore a perfect and beautiful order. It does not matter how insignificant our lives seem, this remains our purpose. Hushpuppy believes she has disrupted the universe by harming her father when she hit him in anger and he fell down. He had left her all alone for several days and had not told her it was because he had been in the hospital. She was fearful of his rage and in over her head in the face of his expectations of how independent she could be as a six-year-old. But he is her dad, and she is responsible to fix what she has broken and care for "who made her." In the journey she has taken, Hushpuppy has gained enough strength to return to care for Wink in his last hours, even though she remains terrified of what will happen to her after his passing.

To do this, she has to face her worst fears. Throughout the film, Hushpuppy has imagined the beasts coming after her. As she returns to be with Wink, they block her way. She turns and stares them down, calling them her friends "sort of." They are her helpers because they have scared her so much that she has been able to call up enough courage to face the reality of her situation. As she sits by Wink's bedside, she offers him food that she brought back from the woman who showed her such kindness. He takes a bite of it and says, "Sweet," and then, in his typical way, "No crying." But they both cry, and the intimacy and love between them is palpable. Upon his death, she personally lights his floating funeral pyre and sends it out to sea, as he has asked her to do.

In the closing scene, Hushpuppy leaves the Bathtub with the last neighbors to evacuate, going into the unknown. Again we hear her voice: "When it all goes quiet behind my eyes, I see everything that made me lying around in invisible pieces. When I look too hard, it goes away. And when it all goes quiet, I see they are right here. I see that I'm a little piece in a big, big universe. And that makes things right."

At the end, Hushpuppy no longer looks scared. She has found a story that empowers her to face whatever happens. She actually appears a bit triumphant, secure in the value of her position as a tiny but important part of a giant universe. Watching her leave the Bathtub to enter a world she had been taught to shun, we know that she will thrive because she is supported by such an empowering worldview.

Hushpuppy's is a Demeter life philosophy and an informing narrative that can help any of us survive our personal tragedies as well as face the immense changes going on in our world today. The heroine's take-away from this lesson is to explore what it is like to feel fully anchored in your own uniqueness and also connected with the people around you, the society in which you live, the earth, and the cosmos. This way you never will be tempted to shut up your inner voice and wisdom, dismiss yourself as an outsider on the periphery, or let other people claim the right to be the norm by which you measure yourself.

~~~~~~

APPLICATION EXERCISE:
A Small but Vital Speck in a Great Universe

Take a few moments to contemplate, moving back into the connected consciousness that you experienced at the end of the previous section if you did that exercise. If not, center in your heart and imagine or feel your connection to the world around you and the grand cosmos. Then ponder these questions, perhaps writing a few notes to help you remember: What loss or losses are you facing or do you fear that you might face? What are your beasts that need to be faced? What is broken around you, and how might you fix it? Once you have answered, become conscious of the story you naturally would tell about these issues, and feel what it is like to hear and believe that story. Finally, imagine yourself with Hushpuppy's philosophy, so that however insignificant you ordinarily might think your life is, you now think of yourself as an essential part of something awe-inspiring and grand out there in which your choices matter to how it evolves. Check in on what that story feels like in your body, around your heart, and in your mind.

Demeter Lesson Three:

Valuing the Generous Heart

WHILE MOST PEOPLE in the United States will say that they believe in equality for women (and for all people), deep-rooted attitudes and structures undermine those stated beliefs, making life more difficult for women than it otherwise would have to be, especially if the way we are treated communicates to us a subtext of being secondary, not primary, in importance. A recent Harvard Business School survey found that while male and female graduates had similarly high career aspirations, the men had a greater ability to fulfill theirs. The chief reasons cited were the intense hours required by businesses of their managers and executives and a discrepancy between male and female expectations about gender roles. The women expected that their husbands would be full partners in their home life. Yet a good percentage of the men saw their own goals as primary and their wives' as secondary and assumed that their wives would carry the major responsibility for child care. The male expectations also prevailed in the workplace, resulting in a reluctance to promote mothers. In addition, it seemed that many of the women acquiesced to their husbands' expectations, perhaps to preserve their marriages.[1]

One sign that we are living in an incomplete revolution is the persistence of outmoded social structures. Business and school hours have remained as they were in most places almost as if social roles had not changed, requiring much scrambling to cover times when children are out of school and both parents are working and commuting. The gap between women's aspirations and attitudes that lag behind *can*, of course, be bridged, once new technological and family norms are in place and social structures catch up with the changes needed to support happy, healthy, well-adjusted families, workers, and citizens. But in the interim, women are faced with any number of signals that the world was not created with us in mind and that if we want to succeed, we simply have to fit into the existing order. And most people assume that "it is what it is."

This is reminiscent of Demeter's experience trying to find out what happened to Kore. Happily, she has an older woman who befriends her, which helps greatly. Hekate suggests that they talk with Apollo, who, as the sun god, looking down from on high, may know more about what transpired. Apollo receives them gladly, happy to explain how things work and that they should not be concerned. Kore's father, Zeus, had agreed that Hades could abduct and then marry his daughter, Apollo reveals, and he proceeds to explain Zeus's reasoning. Essentially, he tells Demeter not to get overly worked up but rather to relax and be content knowing that Zeus has made a good match for Kore. Hearing this, Demeter begins to experience powerless anger over such a decision being made without consulting mother or daughter. She is so furious and depressed that she shuns the company of the gods, choosing to live among mortals. But there is no indication that she actually talks back to Apollo to defend her position. It is as if she shrinks.

From another myth, we know that Rhea, Demeter's mother, allowed Cronus (Demeter's father) to swallow Demeter and all of her siblings with the exception of Zeus, who is Demeter's brother as well as Persephone's father (incest was not considered taboo for the gods, but was for people). Zeus later rescues Demeter and the other gods from their plight. We can see why Demeter might perceive that she did not matter as much as Zeus, prompting her initial response of rather passive suffering in reaction to Zeus's unilateral decision about Persephone's fate. Their mother was so passive that she accepted Cronus's eating their children, saving one only by stealth and never taking her husband on. And since Demeter was rescued by Zeus, she also was indebted to him.

In contemporary communities and workplaces, Demeter women and men keep things going, building community and seeing that the work gets done and people are cared for, while others bask in the limelight, often taking the credit but doing less of the work. At home, Demeter types are attentive to children and other family members and to the care of the domicile. In the workforce, they share these gifts as caretakers and nurturers of people and things. In addition, they often serve as volunteers or activists to assist people who need it most—those who are in poverty, disabled, or discriminated against in some way—or work in the areas of environmental protection, organic farming, or the care and protection of animals.

Like Demeter continuing to keep the crops growing while her daughter remains missing, Demeter women and men provide the daily hard work that keeps our world together, but the downsides are that this work often is taken for granted. Those who do such work often sacrifice much of what they might want trying to make up for a society with a huge care deficit. While the Zeus parts of us seek to shine, our Demeter sides look for ways to help or to work together for an altruistic mission because often they care about others more than their own interests. Women who have internalized messages that they matter less than their male peers, and/or view themselves as lacking options, may suffer in silence when they are objectified or get passed over for promotions they clearly have earned, but that does not mean their pain and outrage are not real.

Moreover, the more what they do is undervalued, the harder they can work to make a difference, both out of altruism and also out of a sense of pride about what they know they have achieved, even if no one else seems to notice. Thomas Merton described this as a kind of modern violence that harms not only those motivated by power lust and greed but also those activists whose altruism takes over to such an extent that they keep helping others even when they themselves are exhausted and depleted, potentially undermining their emotional and physical health. He attributed this idea to the Quaker ecumenicist and philosophy professor Douglas Steere, who said, "To want to help everyone and everything is to succumb to violence. More than that, it is cooperation in violence"[2] against oneself. Thus, if Demeter is dominant in your consciousness, you are as much at risk of overwork as those more like Zeus, and you face the additional danger of forgoing your own authentic path to do what others ask of you. Caring people get satisfaction from helping others, but they also expect that sometimes, at least, others will help them. When their generosity is not reciprocated, they can feel horribly betrayed; in that case, they feel wretched as well as outraged.

But let's now reflect on what the interchange between Apollo, Hekate, and Demeter might mean at a psychological level. In metaphorical terms, we can see Hekate and Apollo as representing Demeter's inner allies. Hekate generally is portrayed as a wise old woman standing at a crossroads, so she is a goddess of the capacity to make wise decisions (what path to take), which Demeter needs to do. As an inner counselor, Hekate would be the deeper, wiser part of Demeter. In addition to being the sun god, Apollo is the god of art, prophecy, and oracular truth. One of his temples near Athens, the Temple at Delphi, was the place where the Delphic Oracle gave advice to the citizens of Athens, advice that was credited with civilizing the city-state. There, her priestesses inhaled fumes from the earth, which augmented their psychic powers, so they could give powerful and helpful counsel to the men and women who came with their questions. Apollo's priests translated what the priestesses said. You may remember from the Introduction that the priestesses originally spoke for themselves before Apollo conquered Gaia's temple. So Apollo would be the equivalent of the inner voice that causes Demeter to doubt herself and her wisdom.

We can think of the two roles—priestess and translator—as going on inside you. Perhaps you have trusted somebody but then been treated by them as unimportant in a manner that takes your breath away and leaves you sputtering. Often, it takes a while to formulate an articulate and appropriate response, especially when someone, Apollo-like, is telling you what you should feel. Women often stop talking when faced with arguments that sound, on their surface, rational and logical, at least for the time in which they live, but something inside them says, "Sounds right. Feels wrong." You may have had this happen. At first you do not know how to reply. Then you feel knowledge percolating up in you, but you cannot yet articulate what you wanted to say. For many women, even today, this process is slowed down by having first to recognize what they really want to say and then to translate that into whatever language we believe is likely to be heard. Frequently, this inner translator has a male voice (like the translating priest), but almost always it reflects the views of those around you in positions of authority.

Here, even Demeter, a goddess, is slow to find her authentic voice to stand up for herself and for Kore, even though she has Hekate by her side. Rather mysteriously, as Demeter loses confidence, Hekate disappears from the story, reappearing only when Kore returns from the underworld. Symbolically, this can mean that Demeter loses her connection with this deeper, wiser voice during this time, hence her despair.

Demeter is not alone in her angry reaction to Zeus's assumption of patriarchal privilege. Greek myths are replete with examples of female gods rebelling against the indignity of gender inequality, including Gaia, the Earth Mother who creates Typhon to subdue Zeus, and the warrior Amazons. Hera, Zeus's wife, births a son by parthenogenesis (without a father), hoping he will challenge Zeus's supremacy, but Hephaestus is born lame and unable to fight his father (although he does become a fine craftsman and eventually the husband of Aphrodite).[3] Greek drama also is full of plots in which men disregard women, leading to the ruin of everyone involved.[4]

The Eleusinian Mysteries tradition presents another story of how a woman wronged creates catastrophe (famine), but it differs from most by demonstrating a path to a positive outcome. When Demeter is wandering, upset, not knowing what to do, she does not yet know that her story will end happily. Her disguising herself as an old peasant mortal signifies how much she has internalized the views of Zeus, Apollo, and many other gods, so likely she is trapped by a conflict between what she believes and what the society and the Olympian hierarchy supports. In Greek drama, the actors wore masks not to conceal who they were but to reveal their inner identity and truth. Thus, we can regard Demeter's disguised state as telling us that she feels like someone much less powerful than she actually is. Being treated repeatedly as less than we are typically becomes a self-fulfilling prophecy that we internalize. Perhaps this is why studies consistently find that women are every bit as competent as men but are less confident and consequently often less successful (see Katty Kay and Claire Shipman, *The Confidence Code*).

Similarly, men can stop listening to their inner feminine wisdom if their society devalues it and thus become less than they could have been. Sophocles's Oedipus trilogy begins with the theme of a Cronus-like father who demanded that his wife kill their son, lest the son kill him, as a prophecy had foretold. She looks at the baby and cannot do it, so she gives him to a servant to solve her problem. The servant carries the baby to a mountaintop and leaves him there, but a shepherd finds and takes pity on him, and takes him to a foreign country, where he is raised by a kindly royal family without knowledge of his parentage. Thus, were it not for kindness, Oedipus would not have lived.

As an adult he returns to his homeland, unknowingly kills his father, with whom he battled over who had the right of way on a road, and after becoming king, marries his mother, though neither is aware of their relationship. When eventually he learns what he has done, in remorse he puts out his eyes and goes into exile. At a literal level (based on beliefs of the ancient Greeks) his story usually is interpreted as being about fate, incest, and inevitable tragedy. Sigmund Freud viewed it as illustrating the challenge that he saw young boys having to face to differentiate from their first love—their mothers, who cared for them—in order to identify and align with their fathers and enter a tougher male domain. Writing in the context of his time, Freud argued that they did this because they feared that if they did not, their fathers would castrate them and they would be powerless, like women.

If the story is read symbolically as wisdom literature and in all three of its parts, Oedipus's killing his father can suggest a rejection of his father's ruthlessness, and marrying his mother can be a way of affirming a more loving Demeter way of being in the world. Freud's reading might help us understand how societal expectations would pressure a boy to act manly and maybe even put down girls as sissies, but Freud was wrong about the body part being maimed: in Oedipus's story, it is his eyes, not his penis, that are at issue.

The actual moral of a story usually comes at its end. As an old man about to die (in *Oedipus at Colonus,* the third play in the sequence), Oedipus rejects his sons, who are off fighting each other for power, but blesses his daughters, clearly stating that he does so because they have cared for him. This demonstrates his eventual realization that love and caring are to be treasured, and that power devoid of care fractures families and communities.[5] In illness or in old age, formerly powerful Zeus types need care, and this often is when they come to value it, though it is too late for them to affect the power dynamics of their time.

Oedipus was oblivious to the identity of his parents and the cruelty of his father, who wanted him killed. By putting out his eyes, he accepts the attitudes of a society that blamed him and not his father for what had happened, even though the outcome was foretold and fated. This trilogy also provides additional context for Zeus's blindness concerning what his actions meant to Demeter. Clearly, damaged sons and devalued daughters were a social problem in ancient Greece that the Eleusinian Mysteries were trying to redress.

Even today, people can see wrongs in the past that they are blind to in the present. For example, not that long ago in American history, under slavery, and subsequently during 100 years of segregation, cruel treatment of African-American men and women was sanctioned by law, by custom, and even from the pulpit. Within any given time, people tend to assume that if everyone else thinks the status quo is okay, then it must be. Even now, when most legal and political barriers to full equality by race or gender have been lifted, many remain

blind to the lingering effects of history on the economic and educational opportunities available to many individuals and to the impact
of an accumulation of microinequities on women and on men from
historically underrepresented groups.

Historical change in gender roles takes time for these reasons, and
because much of what women have accomplished is invisible to policy makers and, to some degree, to the rest of us. This can be traced
partly to the fact that women have been left out of school curricula,
within which males often are the foreground characters, the heroes
that explain our culture and history, and women are the background
or supporting characters. Until recently, written history mainly covered wars and politics—both traditionally male domains. We learn in
school about the male soldiers, politicians, and entrepreneurs who are
credited with having made America great. However, we rarely hear
about the impact of women throughout our history, such as how, from
the 1880s through the 1920s, women who were primarily homemakers
or low-paid workers created the settlement-house movement; gave us
garbage collection, parks, and libraries; and provided services to the
poor. Or how women in the nineteenth century won the right to own
property and led the fight against slavery.

Economic theory generally disregards women's traditional roles,
and our systems of measurement have a similar bias. Evolutionary
economist Hazel Henderson, speaking at the International Conference on Business and Consciousness in 1999, noted that our basic
economic theories are partial because they are based only on men's
economic motivation and participation. Women, historically, and
still today in many places in the world, are employed primarily in the
home, where they work, in Henderson's words, "for love and duty,"
not money. Even in most industrialized countries, women receive less
pay for the same work, but not because their work is less valuable or
needed. Women also do the lion's share of volunteer work. The gross
national product (GNP), the measure economists and policy makers
use as an indicator of prosperity and societal health, is based only on
financial transactions. Henderson posited that women may contribute
more to the health of the world than men do, but most of the work

they perform either does not show up at all or is undervalued in GNP calculations.[6]

In our money-based society, labor that does not involve a transfer of cash or credit is seen as a sideline that can be done in one's spare time. Unreported efforts are invisible, not only to policy makers, but also to employers, and thus the time it takes to carry out these tasks is underestimated. Viewed in terms of GNP, caring for the elderly or teaching our children is regarded as less valuable than marketing drugs to fight anxiety and depression, or manufacturing or selling guns, or producing weapons of mass destruction. Since homemakers are not paid, it has been easy to slip into a situation where a mother will introduce herself as "just" a stay-at-home mom, inadvertently demeaning her worth by the addition of a word.

Although the undervaluing of Demeter work is true for both men and women, this issue hits women hardest simply because care for others still is considered every woman's job, whatever else they do, but particularly for women with children. I strongly resonated to Sheryl Sandberg's point in *Lean In: Women, Work, and the Will to Lead* where she likened men's and women's career advancement to a marathon where people on the sidelines shout encouraging "you can do it" messages to the men, while the women runners hear spectators questioning what they are doing about child care or suggesting that they really do not need to drive themselves so hard. I've certainly felt that myself, as I suspect many women have, so my first response was that this truly nailed the unfairness of women's situations. My second response, which complemented but did not supplant the first, was, yes, but wouldn't the world be saner if the men also were asked, "What about your children?" and "Do you really want to live such a driven life?"

Indeed, the imbalance of Zeus being valued over Demeter in society today creates pressure for both men and women to put self-interest, competition, career advancement, and material consumption before concerns for love, family, and community and before being true to their authentic selves. Men and women naturally internalize society's view of the status of the work they do. They can conclude that jobs that are not well paid are unimportant and try to move to ones with higher prestige and compensation. Yet what would happen if everyone ceased to care about the next generation, or one another? How would businesses thrive? How would society? What would the world be like? If issues of care were ignored completely, the results would be untenable for businesses and the economy, as well as for children, families, and everyone's health and happiness.

Furthermore, Hekate's sudden disappearance from the story conveys another message that I would have missed had someone not asked me to explain it. If Demeter in her mom guise is undervalued, Hekate very often is ignored as if not seen. I remember when my mom was elderly that waiters and salespeople would ask me what she wanted, and I would have to cue them to address her. She complained of having become invisible. The ageism of modern culture means that what wise old women know rarely is filtered back into the collective consciousness of our time. Not being listened to, older women typically focus on various ways to seem younger so that they can continue to be taken seriously.

Recently, I came across Lorena McCourtney's mystery novel *Invisible*, about a woman who reminded me of my mother. The main character begins to realize that as she has aged, people have stopped noticing her. At first she is bummed out about this. As an LOL (her shorthand for "little old lady"), she is very curious about people but does not want to seem to be a stereotypical busybody. However, one day she wakes up in the morning excited by a new story she has told herself: she can use her invisibility to solve mysteries—and that's what she does.

In one episode of the Netflix comedy *Grace and Frankie,* the title characters, older women played by Jane Fonda and Lily Tomlin, realize that the clerk who is supposed to be at the register to take their money doesn't notice them, but an attractive younger woman immediately engages his attention. They leave in disgust, but outside Frankie announces that she has discovered their superpower and puffs on a cigarette from a pack she has stolen, assured that no one had observed the theft because they did not see her. This trickster act helps her recognize that she no longer needs to worry about what other people think of her, but she does need to see herself and discover what she needs to claim her right to live as she wishes at this new juncture in her life.

The generation of gutsy women who spearheaded the women's movement of the late 1960s and the 1970s are now the Hekate generation of our time. Some, like Fonda and Tomlin, still are forging new ground in ways that cannot be missed. Others are openly advocating for various causes, redefining what retirement means, or quietly doing what wise old women always have done: assessing the big picture, evaluating where we are and determining what needs to happen next, then advising young people about what they see. In Hekate, Demeter's connected wisdom adds a more mature dimension to having a resilient consciousness. When we have lived a good long while, if and only if we have stayed alert and alive and not given in to railing against modernity, we can perceive patterns in history and in the human life span. This perspective can help us to forecast and head off dangers in our individual and collective lives— for example, the danger, individually, of squandering precious years in shallow pursuits, or, collectively, of not recognizing how human practices are creating economic and environmental famine for people around the world that is likely to worsen, affecting many more in future generations. Most of all, what this generation knows is that major social change can be achieved.

However undervalued people that express Hekate's and Demeter's worldviews might remain today, the solution to all these problems is for more of us to open our hearts to one another and open our consciousness to recognize our connectedness to the whole of society and of life. We also can voice our true concerns about the state of caring in our countries and in the world without feeling that we have to translate what we are saying into financial terms, and thus reinforce the notion that the only reason to care for people is that doing so promotes profits and social prosperity. It does, but to go there is to let the inner Apollonian priest silence or distort Demeter's authentic voice.

The moral of this part of the story, and a powerful heroine lesson, is to remember to hear and listen to the Demeter and Hekate voices within and without, and to use openhearted, inclusive frames for the stories you tell and live.

~~~~~~

APPLICATION EXERCISE:
Giving Voice and Visibility to Demeter

Over the next few days, use mindfulness to notice when you are reluctant to speak your truth honestly, or to let your truth be seen. Pay attention to what people, causes, values, and ideas you truly care about, and spend a bit of time thinking about how you can give voice to what most matters to you and act in ways that reveal the priorities of your heart. In the process, show appreciation and respect for those around you who are contributing in ways that might be invisible to many and who also might need to be acknowledged in order to gain or retain a sense of their value.

Demeter Lesson Four:

Voting with Your Feet

D EMETER WISDOM OFTEN COMPELS people to leave their
version of Olympus and wander like Demeter, even when
they do not know where they will next find a home. Social
migrations are less obvious when people are not departing from one
region and going to another but instead are leaving relationships, jobs,
and attitudes that are bad for them, for society as a whole, and for the
planet. When people one by one are abandoning houses of worship,
political parties, and friendship networks where they feel that they
are treated as second-class citizens, or where their needs, values, and
interests are not being supported, it may not even be noticed until
someone does a quantitative study and reports it.

As more and more people follow their bliss and live their values,
a great, though largely invisible, migration is taking place. However,
until these trends are observed and measured, the impact of individual
actions is not considered in social policy or business decisions. Nevertheless, such invisible migratory patterns are how unfinished revolutions get sorted out. We vote with our feet by leaving what we do not
want and moving to what we do want, in the process reconfiguring

society. This lesson explores the journey from self-sacrificial caring to valuing oneself as a part of the web of connection, a transition that begins with leaving a setting and/or mind-set and taking an actual or metaphorical journey of some kind.

I cannot help but link Demeter's long period of wandering with the exodus of women from positions in high places, which has begun to be tracked, at least to some extent. Sandberg, along with others, has expressed concern that women who are in line for major executive positions are leaving corporations. But we have a choice whether to see this as a disaster or as a healthy response to what is happening in those organizations. There are a plethora of publications on this subject, but the growing consensus is that the phenomenon is real: many women indeed are leaving good jobs or turning down promotions, especially in organizations where their contributions are undervalued and their family responsibilities are regarded as a handicap.[1] Whether this suggests progress or a setback may differ from situation to situation and woman to woman.

Kathleen Gerson (in *Hard Choices: How Women Decide About Work, Career, and Motherhood*) argues that women are faced with making tough decisions in a context where it is difficult to stay healthy and care for children and other family members, particularly in circumstances where they also are underpaid and their viewpoints not heard and issues not addressed. There is plenty of data from women who have left the corporate world documenting the reasons for their doing so, chief among them gender bias, unequal pay, and not having the flexibility to care properly for their children. The data on why women leave, or never enter, the science, technology, engineering, and math (STEM) disciplines include these factors but focus primarily on the negative impact of macho cultures.[2]

Of course, men also leave jobs when their values are ignored, their strengths undervalued, their opportunities for advancement slight to nonexistent, and they are treated badly. David Allan Coe's 1977 classic country song, "Take This Job and Shove It," remains popular because so many people identify with the fantasy of saying this to a boss and walking out the door, whether or not they actually would do so. My observation is that when women leave jobs, they usually do it very politely, not giving away their unhappiness with a workplace and making sure that they can get a positive reference when they need it—which makes it harder for companies to know what to change to keep them, unless they follow more general research findings.

However, my experience tells me that many people choose to leave because they cannot stomach the heartless practices of their employers. Here are some real-life examples from women (and one man) who have shared their stories with me (I've eliminated any information that would reveal to others who they are). One woman decided to leave her company when the executive team did a cost-benefit analysis and determined that it was cheaper to pay off the families of workers who died or were disabled than to make the factory safer, so they said no to a safety upgrade. Another left when the executives of her firm turned a deaf ear when she alerted them that they were in violation of environmental regulations. Still another woman left after the company gave its executives big bonuses and then called very loyal and hardworking employees into a room, announced that they were being laid off for financial reasons, and had police walk them to their cars. Yet another left because the owner was making millions while she, his second in command, was charged with telling employees that there was no money for any salary increments. A physician left two group health plans because they pressured doctors to see so many patients that it was impossible to do a good job. She feared that if she stayed, she would have patients dying for lack of adequate care. She left and then quickly found a new position in a place whose policies she respected. And the man who left for Demeter reasons did so when his company was hawking nutrition-free snacks in areas of the world where people in poverty have difficulty meeting basic nutritional needs. In all of these situations, the individuals valued their integrity enough to risk losing a very good salary.

In times of economic retrenchment and growing inequality, many women as well as men feel that they cannot afford to leave their jobs, even when they are being treated in dehumanizing ways or are suffering because they cannot be true to themselves. Thus, people often conclude that only privileged women walk out of bad situations. But what I have heard directly from women tells me different, and their courage inspires me. Some examples: A high school girl drops out of the school where she is being bullied by her classmates and her needs are being ignored by her teachers, even though it is at great risk to her future, but gets her GED and eventually goes on to college. A high school graduate first leaves a stressful job at a call center and gets one pressuring poor minorities to take out high-interest loans to go to college. She knows that even if they graduate, they are unlikely to make a decent living, so she leaves again. She then checks out a therapeutic-massage school where she feels suddenly "at home" and enters a field that will not make her rich but will make her happy. A battered woman leaves her home in the dead of night with her two toddlers, not knowing where she can go and how she will support herself. In fact, I have known numerous women who left bad marriages, even though they knew it would greatly lower their standard of living and make raising their children difficult.

The tendency of Demeter types is to remain in very difficult situations as long as they think they can make a difference for the people there, but when they realize that they can't, they go. Similarly, I find that many men and women will stay in an unhappy marriage for the sake of the children, but if they see that the partner or the volatility of the relationship is harming the kids, they take off with them.

It does not take a Nobel Prize level of intelligence to figure out that if a company wants to keep able women employees, it should start listening to the issues women raise and then address them, just as husbands and wives are more likely to stay with their mates if they too listen and act on each other's concerns. But many countries and companies retain the same outlook they had in the past, say, when the people in power were virtually all male, even after the situation changes. The result of the discrepancy between the organizational culture and who works there is that everyone begins to feel stressed and unhappy. It would not be difficult for more of our social institutions to update their attitudes, policies, and structures to match the realities of a diverse workforce and to recognize that healthy families are in our nation's interest. More and more organizations and companies are mustering the courage to let go of their old stories and attitudes, and thus are becoming promised lands for women who simply will not stay in establishments whose values they cannot endorse.

The act of leaving an oppressive or bad situation is itself an archetypal process. The Jewish Seder meal is a ritual celebration of the ancient Hebrews escaping from slavery in Egypt and, by extension, of all the other peoples who have fled oppression. When and if they can, healthy individuals abandon environments where they cannot flourish, moving toward their own vision of the Promised Land, one by one and family by family. All over the world today, people are migrating from places where they are not thriving, trying to find some place where they might.

In the United States, we have a wonderful and long tradition of this. The United States was settled by immigrants who came wanting a better life, some because they were starving, others for the freedom to practice their religion, still others for greater economic opportunity. Our culture is all about holding on to our dreams, even in adversity. When the settlements on the East Coast provided less of a chance to fulfill those dreams, people headed west and settled the frontier.[3]

Years ago, I heard an interview on National Public Radio with novelist Tom Robbins in which he shared a pivotal moment in his life. At the time, he was a newspaper reporter, and as he described it, he was running to the office one Friday afternoon, trying to make it there by 5:00 P.M. to pick up his paycheck, feeling frazzled and stressed. Then he saw a homeless man walking down the street looking blissful, singing joyfully at the top of his lungs. Robbins concluded that this man had "everything" and he had "nothing." So he walked into the newspaper office, picked up his check, and then "called in well"—quitting his job and going on to be a successful author.[4] He didn't quit because the newspaper demeaned him or treated him badly; he left a perfectly good job to pursue his real purpose and gift.

In his novel *Even Cowgirls Get the Blues,* he elaborated: "You've heard of people calling in sick. You may have called in sick a few times yourself. But have you ever thought about calling in well? It'd go like this: You'd get the boss on the line and say, 'Listen, I've been sick ever since I started working here, but today I'm well and I won't be in anymore.' Call in well."[5] Being well, in Robbins's sense, is thriving, not just doing what you have to do to get by. Making that choice releases energy and joy.

There is an interesting paradox in the Eleusinian two-day procession from Athens to Eleusis that is relevant here. Supposedly, it was done in honor of Demeter's wandering in search of her daughter, but it didn't replicate her depressed state. The procession included singing, dancing, and general merriment. Likely, the procession was a model for how to be in transition. So too, our liberation movements typically take to the streets and march. Participating individuals feel invigorated knowing that they are not alone in their discontent and in their desire for equality.

So let's look at when and how Demeter begins to get her energy and vitality back after her leave-taking and wandering. She finally stops to rest, sitting by the seaside at Eleusis, and is treated kindly. Up until this point, Demeter is likely to have been seeing herself in a victim (Kore), villain (Zeus), and rescuer (herself) story. Her sojourn with Queen Metaneira and her family is healing and begins to provide perspective. Their kindness to her penetrates her depression enough that she agrees to drink something, and apparently to eat. The sexual humor of Iambe not only makes her laugh; it undercuts her anxiety about her young daughter being sexual, providing the perspective that such a development is only natural.

In agreeing to be the nursemaid to Demophon, the family's new son, Demeter may be trying to find some outlet for her motherly talents, as well as to show kindness by making him immortal. But of course, in her heart, Demophon could not replace her daughter, and Demeter is by nature a goddess, not a nanny. Some women today leave one workplace to go to another that welcomes them as warmly as the Eleusis royal family greeted Demeter. The new hire then tries to show her gratitude by doing what she can uniquely offer that fits within the organization's clear needs, only to discover, when the powers that be come down on her, that what she is doing may be just right theoretically but is not what they understand or value. That is the time she may decide to start her own enterprise.

Demeter's biggest shift comes when the queen, Metaneira, walks in to discover Demeter putting Demophon into the fire and freaks out. Demeter's immediate response to Metaneira is anger, which often is what is needed to break depression caused by powerlessness. At this point, Demeter is poised to begin to reclaim her power. As she reflects, she also likely sees that Metaneira's horror mirrors her own terror at Kore's abduction, adding perspective to her situation. She recognizes her own story in Metaneira's response, since Metaneira regards Demeter as the villain and herself as the rescuer. Now Demeter can tell herself a different story.

Knowing that what she is doing is not damaging to Demophon, Demeter becomes aware that Zeus also may not mean to harm Kore. He really did think this was a good marriage for her and merely was acting with the arrogance of patriarchs of his time, just as she (with Metaneira) was acting with the arrogance of a goddess. After all, she did not ask Metaneira whether she wanted her son to be immortal. What this means is huge in terms of Demeter's growing narrative intelligence. She begins to shift her story. While she continues to worry about Kore, she changes her goal from getting her daughter and her life back to wanting to see her daughter to know that she is all right and to find out what *she* wants.

I experienced a similar synchronistic turnaround related to the revolutionary story that generally lurks in the background of my consciousness and emerges very specifically when dealing with intransigent authority figures. In my early adulthood, I was the kind of junior faculty member who frequently challenged university administrators on issues of conscience or to push cutting-edge thinking in my field and in academic governance. When I reached an executive level of authority, I continued to push for change, but with a more gradual and, I came to believe, practical, organic approach to doing so. However, I sometimes had to face faculty members who came in full of revolutionary archetype fervor, demanding this and that. I would feel outraged to be pushed up against the wall when I believed

I was doing as much as I could, given the situation I was dealing with.

Then, suddenly, I got tickled, realizing, "Oh, they are me many years ago, and I'm finding out what it was like to be the academic dean way back then, meeting with the angry young me." This awareness—that this was an example of the small karmic events in life—freed me from my annoyance with deans of the past and with faculty members acting just as I had done. The result was that I (usually) could remember what I wanted from such meetings: a good working relationship with these very able teachers and scholars, so that we could work in partnership to improve the institution. I could not have that without listening to and showing empathy for them.

The sequence of events in the home of Queen Metaneira helps Demeter regain her empathy (as a result of the kindness shown her) and her sense of perspective (through seeing the parallel between the queen's upset and her own). She concludes that Metaneira stopped her from making her son immortal not because she was a bad person, but because she was ignorant. Demeter thus decides that she needs to do something about human ignorance, and she commits to creating the Mysteries. Paradoxically, as a result of all this pain and taking the risk of leaving Olympus, she has found a new calling and expanded her horizons.

Let's go back for a moment to Ginette Paris (in *Heartbreak*) explaining how keeping moving after loss or trauma provides us with new experiences that trigger brain plasticity and reconfigure our brains. When we stay in an unhappy situation, we can get very stuck and no longer have an awareness of what else might be possible for us. Not every woman at the Demeter-wandering stage of life physically leaves. Many just begin reading self-help and other literature, going to workshops, traveling, or seeking out new experiences where they are. All these strategies help us to see our lives and situations somewhat differently, so what fits for us changes. This often is the time when we find the match between who we are now and our right home.

At my husband's Aunt Nina's Passover Seder, it is traditional to say, "Next year in Jerusalem," not because anyone at the table is planning to emigrate, but because Jerusalem serves as a traditional symbol of all our personal promised lands. Tradition has it that the ancient Hebrews had to wander in the desert for forty years (meaning as long as it takes) to shed a slave mentality and be ready to live as free men and women. This suggests that what might be most critical for us to leave are the negative, self-defeating stories that keep us stuck. Once we replace them with more empowering ones, we will know whether we need to leave any of the situations we are in or just see them differently.

The point of the journey is how we grow through its experiences. Individuals journey differently, yet some very wise people say that when we are ready for paradise, it does appear. I have learned that this does not mean we will have a life that is carefree and perfect. Rather, we may well find a very satisfying life that provides the next lesson we are ready for—though with better company than we had before.

APPLICATION EXERCISE:
Journeying as a Path to Transformation

Take some time to ponder periods in your life when you have journeyed (physically, emotionally, intellectually, spiritually, etc.) away from one thing and then found something more satisfying, either that you wished for or that was a surprise. Then think about whether observing someone else's behavior triggered you to wonder, "Is that what I act like?" and led you to change. Did recognizing this mirroring allow you to see yourself in a past or present situation differently and perhaps open up options that were not there for you before? If this never has happened to you, stay alert to where you might notice such a reversal of roles in your future and use that opportunity to discover what it is teaching you about yourself. Finally, consider whether you might benefit at this point in your life from letting go of an old story, a relationship, or a situation, transforming your life by transforming the plotline you are telling yourself and perhaps also living.

Demeter Lesson Five:

Standing Up for
What You Care About

A T THE CLIMAX OF ANY GRIPPING STORY, the hero or heroine generally is faced with the most difficult challenge yet. Perseverance at this time is what makes the happy ending possible. If the heroine caves, then the journey continues until she faces her worst fears. Here, Demeter wisdom provides a model for insisting that you matter as much as others. This is an important step in finding win-win solutions that can restore community and your rightful place in it.

The fruit of Demeter's journey up to this point is her shedding the disguise of feeling and acting powerless. She is willing to be seen as the goddess she is and to inhabit her own personal pocket of paradise, her temple. Showing herself in her full glory as a beautiful and powerful goddess, she has asked that a temple be built for her where she can teach people what they need to know to be happy, prosperous, and free of the fear of death. In this way, she has found a physical place to be where she can stop, think, and regroup, and she has claimed her new vocation, which is how many of us recover from loss and difficulty. Eventually, this temple becomes the site of the Eleusinian Mysteries.

From her encounter with welcoming and caring mortals such as Metaneira and her family, Demeter has shifted from the former Olympian story that says that people need to be told what to do and punished if they fail to do it, to a new one that proclaims that while mortals are kind and nice, they are woefully ignorant and just need to learn how things work. She has received a new calling to do something about this. We can expect that what she wants to teach mortals is something she has learned on her own journey. I suspect that her depression is partly due to blaming herself for not being there when Kore was abducted and not being able to say just the right thing to get Apollo's help in getting her back. As most of us do, she likely was telling herself "I should have" stories that implicitly blame her for her daughter remaining in the Underworld. Her challenge is to move from self-recrimination to viewing what she has experienced as an educational process that prepares her for her next step.

Kore is still in the Underworld and becoming Persephone, and Demeter has no way of even knowing how she is and whether she wants to be there or not. Demeter moves into her temple, but still does nothing to make the plants grow. There are several ways to see this.

The unsympathetic regard her as villainous, because the consequences of her actions are serious. But it also could be that, like so many women today, her depression just drained all the life out of her, so she had no energy to keep vegetation alive. It was hard enough just to keep herself going. I see this often with women who are shouldering multiple difficult roles, at home and at work, and constantly are being told or telling themselves that they are not doing enough and not doing it right. When any of us has been mirrored as unworthy and shrink in our estimation of ourselves, it is a wonderful progression when we once again start listening to our authentic inner voices, and then begin speaking and acting in alignment with them. When we do, we suddenly appear to ourselves and to others as if we had grown bigger and more powerful. As with Demeter, this can occur as a result of a series of events: heartwarming kindness (like when the royal family's kindly daughters reached out to Demeter and their ensuing hospitality), an indignity that serves as our last straw (as the queen's tirade was for Demeter), and being in a place where we feel safe (as Demeter must have felt in her temple).

After grief and anger often comes emptiness, a sense of not having any energy. You know how, when you are depressed or discouraged, it is difficult to get out of the chair, yet at the first sign of something that really excites or appeals to you, you are up in a flash? Sometimes, however, the energy drain is all-encompassing, and all you can manage is the minimum needed to eat, sleep, bathe, keep your job, and care for your kids, if you have any. You can see something like this ennui in workplaces where people have given up trying to change things or even just doing a good job. In my consulting practice, I've advised several organizations about how to recover after management did something that so alienated the employees that they shifted from being excited about their jobs and highly productive to an attitude of, "I just work here." This creates a kind of famine, since productivity falls when morale tanks. If this pattern prevails in entire industries, the result can be a depressed economy, because collective prosperity thrives on hope, investment, and spending, which are beyond the capacities of despairing people and ailing companies.

Sometimes this period of depression and emptiness cannot be tolerated any longer. Recuperating in her temple, Demeter also may realize that continuing to make the plants grow is, in effect, colluding with laws and customs that allow fathers to determine who their daughters will marry as well as a host of other ways that men in power make decisions for women without even asking them what they want. At that point, she may comprehend that she is not as powerless as she has been feeling. The only way for Zeus and the other gods to come to understand this is for her to let the famine continue.

Demeter's now consciously being on strike is a mythic prototype of the courageous walkout strikes conducted by workers in the early years of the labor movement, the sit-ins by African Americans during the civil rights movement, the various women's movements across the centuries, the gay rights movement, and others likely pending. They were saying, "We will not be complicit with laws and customs that are oppressive and that dishonor us. We will stand up to power and have the courage to take the consequences of our actions."

These, of course, are macrocosmic events. Our microcosmic actions follow the same pattern but are not necessarily visible enough for the world to know. You likely have experienced times when you decided enough was enough and stood up for yourself and what you believed to be right. It is possible that you are, even now, fed up with acting as if you are less than you are and ready to take courageous action to change the circumstances that have been making you tired or blue.

Demeter, however, wants to be kind and certainly not cause pain and disruption, and she also wants people to see her this way. We need to remember that this is a mythological teaching story, not a literal story, so no one is really dying. However, her willingness to let mortals die underscores a key lesson for us today. She has to act out of character to face her toughest challenge, and she has to become more like Zeus, who has no difficulty causing pain or death to anyone who crosses him.

Women often stay in horrible situations because they know that without them, things could collapse. I recently met a woman who told me her story: She worked for a boss (and company owner) who habitually came in at 10:00 A.M., was then visible at a meeting, then took a long lunch hour, did a bit of management by walking around and chatting, and at about 3:00 P.M., left to play golf or go to the gym, leaving her to do the actual work. She accepted this and worked long and hard because she cared about the company and her fellow employees. Yet when her boss rated her lower on her performance review because she had to take off a couple of days when her children were sick and used this as a reason to deny her a raise, she walked out and left him with the mess. Would the company suffer? Would others working there suffer? Was she entitled to leave? What do you think? You may be glad to learn that as soon as the word was out that she had left, a competitor who knew how competent she was offered her a better-paying position where her work was appreciated.

Demeter does not know how things will turn out when she sticks to her guns as she enters the climax of her story, where things heat up and she will face her sternest test. As you already know, Zeus blames Demeter for creating the famine and enlists others in seeing her this way, too. Zeus sends god after god to plead with Demeter to end the famine, warning that unless she does so, no mortals will be left to worship him and the other gods or to make the requisite sacrifices. Likely, he paints her as a horror of an uncaring woman who would sacrifice all of humanity just to get her way.

Generally, those in power work very hard when challenged by individuals and groups they have harmed or disadvantaged to be sure that others don't see them as bad apples, disruptors, complainers, and generally unpleasant people who should be put in their places. The same thing happens to whistle-blowers, as people in power often fear that their own callous and harmful actions will become known. Even in situations such as I've mentioned where bosses know that women or men are quietly leaving because of their company's or even their own deplorable practices, they generally require the person who is quitting to sign an agreement that they will not disclose anything about them or the company in order to get even the small severance package they are owed and truly need to tide them over until they find a new job. This behavior of those in power at any level is just a natural human defensive posture to keep from being found out.

But even in the light of such continuous pressure and having her reputation tarnished, Demeter remains steadfast, basically saying, "No one eats until my daughter stands before me and I know she is all right." Her modeling provides a very important lesson for women today. No matter how much we want to be nice and caring and to be seen that way, sometimes heroines just have to be tough in the service of whom and what they love.

When he realizes that she will not give in to pressure, Zeus relents because she has won the power standoff, and he recognizes how naïve he was to think his power over social organization was greater than hers over nature. Demeter and her daughter have a joyous reunion and Hekate returns after her mysterious absence.

When Persephone returns, she seems not only unharmed but also cheerful, which makes it easier for Demeter to accept the seasonal compromise (of Persephone being with her mother for part of the year and with Hades for the rest). When Zeus invites Demeter and Persephone back into the family of the Olympian gods, with Persephone remaining an Underworld goddess as well, Demeter agrees, as does Persephone, reestablishing a sense of community, which is a happy ending for the gods and humankind. It also reflects maturity and growth on Zeus's part. Demeter and Persephone then run an organization (the Mysteries)—like modern women starting a business together. In all the visual depictions that remain at Eleusis, both look quite happy, and Persephone also looks contented in the Underworld.

Moving for a moment to thinking about this ending in the context of a mother-daughter love story, we can conclude that Demeter, like so many of us, made the transition from being a worried and potentially controlling mother to being a friend and colleague of her daughter, who grew up into a powerful goddess. This is an important transition for all mothers and daughters (and for parents and children more generally). The scene upon Persephone's return of a mother and grown daughter catching up with one another suggests that they are making this transition. If you are the mother or father of a grown child, you may have found ways to recognize and possibly even ritualize this transition from childhood to adulthood with your son or daughter, perhaps at the time of their high school or college graduation, and know how satisfying it can be. I remember fondly being with my eighteen-year-old daughter on a rooftop in Athens, spontaneously inventing such a ritual as we talked about the coming change in our relationship and then stepped over a threshold arm in arm.

Throughout her story (and in all its levels), Demeter demonstrated how independence could support interdependence. Taking a difficult and lonely stand eventually folded her back into the web of human-god connection—but now as an equal who had to be treated by Zeus and the other gods with respect, not as a lesser goddess whose contributions could be taken for granted. The moral of this story is that we can do likewise. As more and more women withdraw their energy from shoring up what they do not want and take a stand for what they do want, the unfinished revolution comes closer to being completed. We can reinforce such progress by spreading the news of where bits and pieces of it already have been realized.

There are pockets that preview paradise everywhere in the world. At the end of part 2, the Zeus chapters, I will include examples of gender partnership and its results in organizations, but it seems better to end part 1, the Demeter chapters, with a larger societal example. A recent study of women in leadership roles in various places in the world came back with the astounding finding that women leaders in Sweden are *not* chronically exhausted, seem extremely happy, and report satisfaction with work-life balance. Why is this so? Sweden adopted social policies that both encourage women to work and enable them to have and raise well-adjusted, well-educated children while also taking time for themselves. Adequate maternity and paternity time is mandated, preschool is subsidized, and generous time for family vacations and resistance to working hours that undercut employee and family health are national policies. According to this study, men are happier, too, because they have reasonable hours and get to spend real time with those they love. Companies do better, as well, since employees are more efficient, effective, and less ornery because they are less stressed.

When the results of this study were reported at a session of an International Leadership Association meeting I attended, they created a euphoric uproar in a room full of amazed professionals from various countries, most of them women, energized by hearing that such outcomes not only were possible but actually were happening. Hands flew into the air as people asked questions, wanting to be sure the results truly were accurate. A discussion ensued about how to achieve similar outcomes elsewhere. The conclusion was that this could be the case in the United States and other countries if the following questions were on the national agenda: Since women have entered the workforce in large numbers, what needs to change to make this development positive for women, men, children, businesses, and societies? How could all children be loved, educated, and well cared for? How can businesses retain hard-working, capable, and talented employees?[1]

Happy societies encourage self-reliance, ambition, and personal freedoms while also promoting care for the quality of life of their citizens, and they revise their narratives when the world changes, which may be why Scandinavian countries come out at the top of the United Nations World Happiness Index.[2] They also provide Zeus-Demeter balance by adopting social policies that protect competition and individual rights while allowing more time for family life and creating a stronger safety net. In my country, the United States, we probably would have accomplished this already were it not for cultural narratives that hold us back. Two that come to mind are the myths (in the sense of falsehoods with no actual data to support them) that market forces will take care of all problems, and that policies that mandate care for people represent "godless communism" or "socialism," a term often used without distinguishing between its forms in democracies and dictatorships. There are people who make their living demonizing words (like *liberal, feminist,* or *the 1%*) so that new terms continually have to be invented for the same group or set of beliefs. Narrative intelligence can help us stay conscious and recognize when we are being manipulated to dislike code words for ideas we might support if their meanings remained clear or might critique with greater clarity if we do not. And the critical thinking narrative intelligence fosters also can help us to recognize when we are being encouraged to stereotype members of any group rather than treat them as individuals.

Contemporary heroines can be like sleuths in mystery stories, but in this case solving the mystery of who are those actually maiming and killing Demeter values and what are their motivations so that their activities can be revealed and curtailed. However, since many women have too much, not too little Demeter, it is time to move on to explore the other archetypes that are equally important if we are to achieve the Eleusinian promise of happiness, prosperity, and freedom from the fear that holds us back from experiencing our greatest good.

APPLICATION EXERCISE:

Imagining and Reinforcing Pockets of Paradise

Reflect a bit on where you have experienced, or are experiencing, a pocket of paradise (POP) that you would be willing to stand up for and protect. What is your vision of an optimal POP? Once you have identified it, find a way to keep that vision in your heart and mind. You might write a brief description of it or draw a picture or symbol or create a collage that evokes it for you. Put this document someplace where it can remind you of what such a paradise would be for you, so that you can notice elements of it that come your way and weave them together as you create the tapestry of your life.

Then contemplate how brave you are willing to be in the service of fostering such a good life for yourself and those you influence. As a follow-up mindful practice, be conscious of the stories you tell yourself that devalue your POP and reframe them into narratives that move you closer to realizing your vision.

CAPSTONE EXERCISE:

Dialogue with Demeter

In a journal or notebook, engage in a dialogue with Demeter where you speak as yourself and the Demeter of your imagination speaks for her. Start by thanking Demeter for the ways she has enhanced your life and then ask how she would like to be expressed in you. You can then say what you would like from her, telling her how you think she would be helpful but also saying how she might not. Allow a natural dialogue to develop through which you come to an agreement about Demeter's future role in your life. End on a note of appreciation.

Part Two

~~~~~~

ZEUS

Zeus and the Way of Power

THE ZEUS ARCHETYPE is important for any of us who wish to realize our potential. It also is central for the success and well-being of social organizations. For families, workplaces, and government to function well, we need to claim our power to analyze what must be done, make decisions, and act with energy and effectiveness. In the family, if parents do not take responsibility for providing for and raising their children, the next generation won't turn out very well. In businesses, if workers aren't good at what they do and don't constantly aspire to improve, they and the organization will not succeed. If, in a democracy, citizens do not stay informed and active, their interests won't be served, and important shared dreams may never be realized. Zeus's story helps us to recognize that shared power is a good thing, but it works only if we do our parts. That requires us to develop our own capacities to support the good of the whole.

In all times and places, group prosperity has depended on individuals developing their skills and abilities so that they could make tangible contributions to their communities. Often, it was expected that most people would carry on a traditional family vocation, or if that did not work for them, apprentice with a neighbor to learn another needed trade. In many indigenous communities, adolescents have been sent on vision quests to discover their vocations. While it has long been the case that some individuals in modern Western cultures feel a calling to ministry or to the arts, most people have focused mainly on finding jobs that fit their abilities and earn them a good living. In more recent times, however, more individuals have aspired to find work that is related to a sense of purpose aligned with their abilities.

We live in a competitive society that demands that we hone our skills if we wish to succeed, and it is primarily through striving and accomplishment that we develop strong and healthy egos. Ego strength allows us to establish clear boundaries, stand up for ourselves when conflicts occur, and achieve the independence needed to manage our own households and professional and financial affairs. Although women still are expected to play nicer than men, we have to be perceived as competent and tough to succeed, and if we want to be true to ourselves, we have to be willing to fight for our values and our rights.

To challenge Zeus's decision to allow Hades to abduct and marry Kore, Demeter had to become more like him and take a stance for what mattered to her even if it had consequences for others. In our own time, very caring and generous people who embody Demeter's primary gifts may have difficulty focusing enough on their own self-interest to do well materially or to meet other personal goals without having access to what the inner Zeus can provide for balance.

Just as Zeus is the god responsible for social organization in the public sphere, women entering traditionally male domains are better prepared to thrive when they understand Zeus's mind-set and tap into his strengths. Access to Zeus becomes even more critical for women working in partnership with men in the home, at work, and in the community and as we become more and more successful in our careers or civic activities and as our responsibilities and our zones of influence grow—sometimes in big leaps (as in a major promotion or starting a new business).

The Zeus archetype currently is morphing in significant ways, and not just because women are entering realms that in the past have been considered male domains. The male advantage of greater physical strength and men's superiority in objective and logical thinking, which arose during periods when women were not educated in those ways, are things of the past. Machines now can do most of the world's heavy lifting, and computers are able to "think" without their feelings getting in the way. In places where women have access to mental and physical education and affordable and safe contraception, the primary factors holding us back have been eliminated.

Together, these changes require men to evolve beyond macho expressions of the Zeus archetype to reframe masculinity in the context of gender partnership, just as new opportunities challenge women to integrate Zeus virtues into our understanding of what it means to be a woman today.

The Zeus Archetype

Zeus is an expression of the archetype of the king or queen, or, in modern times, of the CEO or president. The Zeus part of us enables us to claim our power and to exercise our responsibility for family, workplace, civic engagement, and social organization—that is, to make sure that all parts of our world run well. Zeus provides the awareness that all of us are the CEOs of our own lives and need to be in charge of and able to manage our own affairs.

A model for exuding confidence, Zeus is portrayed in Greek mythology as charismatic and charming, and he runs a tight ship. He is the ruler of the earth and the sky, and as a sky god, he has the advantage of seeing the big picture. This allows him to learn, over time, to rule with a systems perspective, understanding how the parts fit into the whole. An embodiment of power, he often is depicted hurling lightning bolts down from the sky. You can see him today in parents who can say with authority, "Go to your room," and the kids stop misbehaving and scurry to get there without the need for threats, or the boss who has no trouble presenting a goal, making clear who is doing what, and getting compliance. We all (even children) need a bit of Zeus, or else we are likely to be pushed around by bullies.

A good dose of Zeus energy helps you set strong boundaries, have the courage to defend yourself, and be successful in your career and life. An awakened inner Zeus empowers entrepreneurs to create new and successful businesses, politicians to run for office and win, leaders to create and sell a vision for the future, and any of us to clarify our ambitions and achieve them. By choosing to value this archetype and learn from it, you can gain an inner ally who helps you build your strengths, prove your worth, and get the attention you deserve—all while safeguarding your individuality. If that were not enough, it enables you to stand up for yourself when you encounter gender or other biases at home, at work, or in the public sphere.

As with all the archetypal characters in this book, the inner Zeus can be expressed in different times and places. The images of a rather autocratic Zeus originated in times when rulers were expected to govern with a very heavy hand and dole out punishments to keep people in line. The shadow of his archetype is revealed in the behaviors Zeus initially demonstrates that establish his power through violence and harsh punishments. The negative view many of us have about power comes from observing or being hurt by the more negative Zeus qualities that persist in our culture in the bullies, the autocrats, the unrepentant Scrooges, the dictators, the torturers, and the Francis Underwoods (the cynical, manipulative politician in *House of Cards*) of the world.

Although the Zeus we meet in mythology has some behaviors that now would be deemed inappropriate or even cruel, he is always learning, and thus is a model for how we can learn and evolve rather than holding on tight to ways of being that keep us from addressing contemporary problems. His growth within the Eleusinian narrative that begins this book parallels the shift in consciousness in an Athens that then was moving toward a great, though limited, experiment in democracy. With regard to our time, the prospect of famine that Zeus had to confront prefigures scientific findings that create new leadership challenges. Humankind cannot conquer nature as if we were separate from it, as prior cultural narratives told us we could. The threat posed by climate change reminds us that we are part of nature, so what we do can unbalance the ecological system with disastrous effects. However, science, and consequently our cultural stories, is evolving rapidly, so that it is possible for us to analyze what is happening and perhaps head it off. It is the Zeus archetype that can help develop the social consensus and political will needed to translate this new knowledge into public policy.

The Zeus Story

As I pieced together Zeus's story from various myths about him, I saw that he went through a learning cycle that can help us thrive in these challenging times, but we have to interpret it metaphorically to understand its meaning. Here is the short form, which the lessons that follow will flesh out in greater detail: Just as many of our personal issues can be traced back to some trauma in our family histories, the deep background for the Eleusinian Mysteries is the story of Cronus, who swallows all of his children except Zeus. In Greek mythology, Cronus was the king of the Titans, the race of gods that ruled before the Olympians. Fearful of a prophecy that a son would replace him, Cronus swallowed five of his six children as soon as they were born. To do something this horrible, he had to be taken over (we could say swallowed) by a power complex, where maintaining his power was more important to him than his children.

When Zeus is born, his long-suffering mother, Rhea, the original Earth Mother, finally has had enough. We can imagine her pain in losing so many children and applaud when she has the courage to save Zeus. She wraps a stone in swaddling clothes and gives it to Cronus to eat, then sends Zeus to Crete to hide. There, he lives alone in a cave, and we can imagine that he must have felt both rejected by his father and abandoned by his mother.

However depriving his start in life, Zeus manages to succeed. He has the foresight to learn all he can about the wisdom of the advanced and egalitarian Cretan society, which later serves him well. Zeus decides that he will replace Cronus as the chief god and rule in a more just way than his father. When he grows up, he returns home and rescues his siblings, with some help from others, by feeding Cronus an herb that makes him regurgitate them. They join Zeus in a rebellion against Cronus.

Many of the gods and goddesses must have felt oppressed by Cronus's rule, and they help Zeus overthrow his father. However, others are attached to how things were done when Cronus was in power, and the status they received from their affiliation with him, so they fight against the rebellion. Nevertheless, Zeus proves his strength and fitness to rule as he leads the rebel forces to victory, and generally he is a better and wiser monarch than his father. He marries a Titan— Hera—to make peace with those he vanquished, but the marriage is not a happy one. He engages in numerous infidelities, and Hera often is portrayed as weeping or wildly jealous and punishing. Zeus's daughter, Kore, is the product of a roll in the fields with Demeter, but there is no record of his being an involved father until Hades asks for his blessing to marry her and Zeus agrees. However, when faced with a famine that threatens to wipe out humankind and deprive the gods of the sacrifices that are essential to their welfare, he experiences a huge wake-up call, recognizing that his management style needs to accommodate many more variables than it had before.

Chastened by this realization, he sends Hermes to bring his daughter, now called Persephone, back to the Upperworld, invites Persephone and Demeter back into the fold of the Olympian gods, and wholeheartedly supports the creation of the Eleusinian Mysteries, for he wants humans to understand what he has learned. But most of all, he wants to ensure the long-term fertility of the land as the key to prosperity, success, and happiness for mortals and gods.

Within his story, we can trace political evolution from aristocracy (Zeus's father), to meritocracy (Zeus earned his way, starting as the cast-off son), to greater democracy, with the recognition that all the gods and goddesses should have a voice in their own and in human affairs. Instead of thinking that he is superior because his job is to run things, Zeus comes to understand that his is just one of many important jobs that need to be done.

Cronus embodies the primal undertow of the Zeus archetype—frequently described as a power complex, where the need to have power over others possesses us, often unconsciously. Cronus's swallowing of his children can be interpreted as a metaphor for parents—or anyone in a position to control others—who declare what their children or others in their charge should think and do and who compel obedience with fear. Cronus is the archetypal presence within all repressive regimes. If everyone in your company is scared of doing anything to displease the boss, his energy is present. Recognizing Cronus at work builds your capacity to spot the power complex story when it emerges in you or around you, so that you can avoid the toll it takes on everyone involved. It also explores how inevitably we are swallowed by the system in which we live, and how to recognize and escape the shadow elements within a meritocracy.

The following lessons trace Zeus's story to illustrate how the development of a healthy ego structure can complement the wisdom of the heart. The lessons walk you through the process of deconstructing any stories that are keeping you locked into limiting patterns, so that you can differentiate more completely from others, claim your authentic strengths, and follow your genuine interests. They also will help the Zeus part of you develop healthy boundaries and clear goals, take responsibility for your mistakes or limitations without having to deflect blame onto others, have the resilience to utilize failure as an opportunity to regroup and thrive at a new level, and no longer have to be one up on anyone to feel secure in your identity and your world.

In addition, the lessons parallel the evolution from a meritocracy to a truer balance with democracy that is now under way but still under siege. All the liberation movements of our age, along with the increased expectation that people will be treated as individuals who deserve fulfillment, not resources simply to be used, are moving us in this direction. At the same time, however, the growing influence of the few who have accumulated economic power and seek to maintain and enhance their advantages is slowing that progression substantially. Right now, Western democracies are stuck with competing stories:

one that sees economic prosperity as the inevitable product of a contest that determines winners and losers, and another that holds out a promise that even greater abundance results when each individual can fulfill his or her potential and thus contribute to the whole. In Zeus's story, each of these narratives dominates in a stage of his journey.

In my own life, as I moved into parenting and administrative jobs, I had to jump-start Zeus, since this archetype had been deemphasized in comparison with the Demeter values encouraged in my rather kindly and gentle upbringing, which left me too trusting of others. In all of these roles, it was important for me to trust but verify, learning through the school of hard knocks to be realistic. It took me a while to recognize how many avarice-motivated people are out there, often camouflaging their greed behind a virtuous and caring affect or social policies designed to benefit them by exploiting people's fears and prejudices. The Zeus archetype can help us recognize these dangers and protect against them, because they are the negative pole of his own nature. We cannot see what is unconscious within us.

But beyond becoming less naïve, I needed to develop the many very real Zeus strengths that could help me be successful, especially in my work life. My husband, David, coached me, encouraging me to "lean in" way before Sheryl Sandberg made that phrase famous, and I encouraged him to lighten up a bit as a dad. An early mentor who was the dean of the college when I was a young professor and administrator was terrific at telling stories that helped me, and others, understand the academic world and how it worked in practice.

In my consulting and coaching, I found that women who had experienced autocratic fathers, mothers, teachers, or bosses often had difficulty claiming their authority. At some point, they had told themselves a story that said, "I hate being hurt like this, so I'll never act in this bossy way," which led them to suppress not only the negative side of the archetype, but also the whole of its confidence and willingness to take command and exercise leadership. This inner narrative would become troublesome when they were promoted into management at work but were seen as resistant to lead, or when as mothers they could not crank up the energy to tell their children what to do with enough authority to be obeyed. Equating power with its shadow forms causes many of us to repress our self-assurance to the extent that we become powerless and let the bullies walk all over us.

The solution is to awaken the Zeus archetype within us to toughen us up, but also to balance it with Demeter's connected consciousness so that we remain focused on the greater good. While most women can benefit from Zeus's ability to put his own interests first, the shadow side of Zeus does this in a way that is either oblivious to what it means for others or actively ruthless. The heroine's path balances self-interest with the awareness of how interdependent people are. For example, in our modern economy, very few of us are exempt from the impact of a major recession. If climate change proceeds unabated and more "natural" disasters keep occurring, the challenge to our thriving will escalate greatly. Moreover, in a globally interconnected world, we all are affected by the suffering of others, by related depression and anxiety, and by international threats such as terrorism and disease. You can consider this book as being like homeopathic medicine, where the evolved manifestation of the Zeus archetype is the cure for the damage still being caused by its less advanced expressions.

Zeus is the god of power, and we all need power in its various forms—*personal power, power to, power for,* and *power with*—which can help us build our own capacities, achieve what we desire, exert courageous effort for the greater good, and be strong and collaborative members of a team. Some have a primary calling to lead or manage and need higher-level Zeus skills than others, but we all need elements of Zeus's capabilities just to take charge of our own lives. The lessons that follow explore how you can

+ overcome the fear and anxiety that fuels a driven life;

+ declare your independence from living by others' rules;

+ unleash your passion and focus your actions;

+ regroup and rethink as you begin to know more; and

+ move from *power over* to *power with* relationships.

The Zeus Archetype
Mythological Zeus: chief god, god of Olympian deities, sky god
Primary Heroine Lesson: the ability to take charge of one's life and affairs
Narrative Progression: moves from anxious arrogance to grateful collaboration
Gifts: strength, discipline, determination, confidence, and strategic thinking
Historical Gender Association: with the masculine in men and women
Decision-Making Mode: trusts gut feelings and strategic intelligence
Inner Capacity Developed: a strong, healthy ego
Counterproductive Forms: ruthless tyrant, entitled autocrat, narcissist

Zeus and the Eleusinian Promise
Primary Happiness Practice: having a sense of purpose and commitment to goals
Contribution to Prosperity: aspiration, planning, perseverance, and hard work
Contribution to Freedom: building strength, confidence, and determination

APPLICATION QUESTIONS:

Based on what you know so far about Zeus,

Do you have too much, too little, or not enough of this archetype?

If it is present in your life, was it a family legacy (i.e., you were taught to be like that and are), a vocational ally, or something deeply and authentically you?

Do you like people who reflect Zeus's stance in the world? Why or why not?

Zeus Lesson One:

Overcoming the Fear
That Fuels a Driven Life

S OME PEOPLE WHO SUCCEED in reaching the top in their field
become so attached to remaining there that their fear of losing
their power, status, and money ties them in knots. The same can
happen to experts and other opinion leaders, as well as to business and
political leaders. And it can happen to any of us who find ourselves
holding on tightly to any status or economic advantage, fearful of sliding down the economic ladder. Instead of enjoying what we have and
taking pride in our achievements, we can become fearful, clinging to
or attempting to expand whatever power we have.

The first step in avoiding this is to understand it as a power complex kicking in, which is the subject of this lesson. Any archetype
becomes a complex when it possesses us, generally in its shadow form.
For example, Demeter's care complex can take us over with the result
that we become controlling in a codependent way. (In this, she potentially also could emulate Cronus, who was her father too.)

When the power complex grabs hold of us, as it easily can, we may
feel that whatever level of status we have attained is now essential, not

only to our sense of who we are, but also to the survival of whatever enterprise we are engaged in. Recognizing this can help you understand the behaviors of people when they do not deal well with losing control over children (who rebel or grow up), boards or bosses (who fire them), employees (who do not perform or take another job), customers (who choose another product), peers (who exclude them), or the populace (who elect someone else). This loss of control can result in disorientation, depression, or desperate acts to regain it. Even worse, dictators may torture and execute others to keep their power. Most of us, under the influence of Cronus, just torture ourselves.

What you need to know to enter the mysteries of power, and protect against their drawbacks, is revealed in ancient myth and literature. Sir James George Frazer's seminal book on mythology, *The Golden Bough*, explores myths that portray the king's reign as tenuous and contingent on his connection to the land and on his physical prowess. In the myth of The Golden Bough, the king had to guard a certain limb of a tree, which was the symbol of his physical strength and fertility. If a challenger tried to steal it, a fight to the death would ensue, with the winner becoming, or remaining, king. Consequently, the king was constantly anxious and afraid even to sleep lest someone sneak up on that tree.

Other myths throughout the world link a king's continuance in his role with his ability to demonstrate his strength in battle and with his sexual capacity, providing context for Zeus's rather off-putting aggressive and sexual exploits. The former shows that the king can protect the community from outside harm, and the latter comes from an ancient belief that the fecundity of the land depends on the virility of the king. Saul Bellow's entertaining novel *Henderson the Rain King* offers a peek into what it might be like to be a king subject to such expectations. In it, a fictional African tribal king explains to the novel's protagonist that his life will be sacrificed when he no longer can satisfy all the women in his harem and produce more children. While we may not currently expect this of our leaders (preferring them to have monogamous marriages and be exemplars of family values), macho emotional attitudes live on in some locker rooms (and workplaces), with tales of sexual exploits and aggressive wins. This feeling persists in the more general anxiety many of us experience when our responsibilities are so great that we fear we cannot sustain them over time.

Yet the negative impact of Cronus is reflected differently in different times in history and by different social and economic classes. Historically, the anxious alpha-male pattern showed up again in feudal lords and ladies (ruling over serfs), in kings and queens, and in aristocrats more generally who maintained that their superiority was innate because it was inherited. In the plot of stories like that of Robin Hood, the hero overthrows an evil ruling class (King John) that has been robbing from the poor so that the wealthy can hog even more wealth, luxury, and power than before. The hero then helps to reestablish a government (in this case, led by King Richard) that cares for the people and creates a more prosperous and harmonious kingdom.

We can observe a similar phenomenon in our own time, in our economy and our politics, in the people who increasingly run things by virtue of their wealth and consequent power. The persistence of their children's advantages in resources, connections, and access to education creates a virtual aristocracy, and their investments grow their fortunes exponentially. At its worst, the family line obsessively swallows more and more of the available resources and then hoards them, contributing to greater and greater economic inequality.

According to thought as ancient as Homer's and continuing today, the moral duty of aristocrats, and the way they can escape being swallowed by less-desirable Cronus qualities, is to recognize *noblesse oblige,* which essentially means, "To whom much is given, much is required." That this is a French phrase reminds us that the French Revolution occurred because the aristocracy was neglecting this responsibility. Even an alpha-male primate does his duty, which is role defined in nature, but humans have free will. Some of our more enlightened wealthy citizens now use their fortunes to try to address social issues and health problems and to give the less fortunate a leg up. Collectively, the world's richest billionaires have amassed such a large percentage of the global GNP that together they could solve many of the world's current problems if they just would. For example, a 2014 report from Oxfam, *Working for the Few: Political Capture and Economic Inequality*, found that "Almost half of the world's wealth is now owned by just one percent of the population."[1]

In addition to changes in tax, inheritance, and other policies by governments, addressing this discrepancy would require a shift in the stories many of the wealthy tell themselves, and from which they derive self-esteem, from "The richer I am, the better I am" to "The more generous I am, the better I am." Such a narrative modification would entail balancing Zeus's power with Demeter's generosity.

However, it would be a mistake to leave the solution to the world's problems only to those who are the most cut off from awareness of the extent of human suffering and environmental degradation. Overcoming Cronus is an inside job for all of us. In early tribal human life, when survival often was in question, each person who had a skill or ability that contributed to the success of the group would have stature and importance. Today, the issues before us are too great to assume that only the few who are considered to be "the best and the brightest" or our governmental leaders can solve them. Instead, we all need to become the best we can be and contribute what we can to creating a more just and prosperous society.

Living as we are in what is conceived to be a meritocracy, where anyone (theoretically) can succeed by virtue of hard work and competence, the Zeus archetype shows up in men and women who become the monarchs of their own lives and who prove their ability to lead or influence others by virtue of their achievements. The downside of a meritocracy is that it can foster an attitude of valuing only the people who make it to the top, while blaming those who fail to do so for their plight, which is seen as a natural consequence of ineptitude or laziness. Such a mind-set can cause the winners to feel entitled, those in the middle to sacrifice their authentic journeys to constant striving to achieve more and more, and those who struggle financially to accept powerlessness and poverty as their due.

Yet ideally, a competitive, capitalist society should motivate each of us to hone our natural strengths, so that we make a valuable contribution to the whole and have the resources we need to realize our potential and enjoy our lives. However, if we believe that our worth comes from where we are on the ladder of a socioeconomic hierarchy, then we are going to be driven continually to attain more status, which often is equated with more money and power. Such economic aspirations begin to consume us, just as Cronus consumed his offspring, and we will devalue or disempower anyone or anything that threatens our continued swallowing of the world's resources. When we cling to these artificial measures of our worth instead of following our evolving purpose, we will begin to feel empty, always needing more to fill that hole where our real self should be.

If we lack awareness of our innate values and strengths, we may choose a mate in order to enhance our status in the same way someone might acquire a house or car to impress, rather than because it serves their current needs and matches their aesthetic sense, and thus pleases them. This can lead to a sense of dissatisfaction and emptiness that some deal with by trying to fill it in other ways, such as by ravenously desiring more—more money; more houses; a new, upgraded spouse; more intense experiences; and so on. I've learned that people who achieve high status, riches, and fame can feel as lost and empty as the rest of us sometimes do—maybe even more so.

An early wake-up call for me came shortly after I had graduated from college, was newly married, had found an apartment with a pool, and was teaching high school English, doing pretty much everything I was expected to do and even knew to want. I thought all was well until one day I went home, baked an oatmeal cake, and devoured most of it. Even that did not slake the emptiness I felt. To assuage that hunger, which actually was not for food, I had to begin to know who I was and what I wanted to do and be. I'm grateful that I got the message relatively early in life. As Joseph Campbell famously warned, many people do everything they can to climb the ladder of success, but once they get to the top, they realize that it is up against the wrong wall. It is never too late, however, to stop, regroup, and get in touch with who you really are and what your life ultimately is about.

The Zeus archetype is evolving along with human patterns of thinking. We are emerging out of a time when people looked at nature and observed "the survival of the fittest," and then took this as a metaphor for how capitalism should work, concluding that the key to organizational success would come from having leaders who fought their way to the top, perhaps even in a ruthless way. As your success narrative shifts, the goal is to find, not the one right way, but the right way for you to be the best you that you can be. Scientists now perceive the need for balance in ecological systems, and economists are investigating what global financial interdependence means for the way the economy works and how growing income inequality decreases the prosperity of the entire society.

At the same time, many in the United States are beginning to realize the urgency of more fully embodying the egalitarian promise of democracy to better balance the meritocratic ideal. This can lessen the drawbacks of each. At the individual level, the belief that our worth comes from our position in the social hierarchy, and the anxiety this engenders, can leave us feeling swallowed by the sense that we never have enough time to do everything we should be doing to live up to the standards of excellence in our society. During antiquity, according to Plutarch, the Greeks believed that Cronus was an allegorical name for Chronus, the god of time. The story of Cronus eating his children can be interpreted as representing the way the past sometimes consumes the future, how the older generation suppresses the next, and how few of us have enough hours to get done everything we want to do, whether in a day or in a lifetime.

We all are swallowed by linear time, which creates an awareness of the inevitability of our mortality along with a desire to be eternal. Out of the desire to escape mortality comes the urge, for some at least, to create a business, a building, a monument, or a timeless piece of art with the idea that something of us can live on. Our genes live on in our children, if we have them, and our ideas in the people we influence. Most of us want our children to be happy, but we also wish they would retain our habits and beliefs. This is not just because we think these constitute the best way to live but also because when others retain them, they carry a bit of us along, too.

When the Zeus archetype is active in us, the desire for control can engender a desire to stop the flow of time and change. The solution to this eventually comes in finding a good balance with Demeter, Persephone, and Dionysus wisdom, but it begins with recognizing the dangers within the Zeus archetype itself. What we do not have a name for can easily take us over without our noticing why we feel so stressed.

In our personal or work lives, we may harbor a fantasy that we can pass the torch along to others and they will not change anything. You may have witnessed this in the struggles of founders of programs and organizations who remain in place even when they no longer actually want to continue with that level of responsibility. In my own case, at those times when Zeus was awakening in me, I kept getting drafted into administrative leadership positions. On several occasions, I was the founder of a program. In one such instance, I left a position as director of an academic program to become the academic dean of another institution and spent days writing up notes about everything my successor should do. Later, I realized that she never even read them—but somehow did a great job anyway! In two later situations, I succeeded a founder who was still powerful and around, and my previous experience helped me empathize with how hard it is to let go of what you built.

In families, we now have a generation of helicopter parents who swoop in to rescue their children long after they should be making their own way, an issue that causes chagrin in many bosses and human resources directors, who wonder why a mother is calling about the performance evaluation results for one of their grown employees. It is natural to want one's legacy to go forward and, consciously or unconsciously, to be anxious about a future that evolves differently than you think is desirable. Yet this is how life works, and accepting this archetypal pattern can help us achieve a sense of personal worth deeper than our roles and accomplishments, so that we truly can enjoy and experience where we are without so much fear for the future. The younger we are when we come to terms with this, the happier we will be.

Linear time also is scarce. We all have only twenty-four hours in a day, and for many of us, there never seems to be enough time to do everything we would like to do or even what we are responsible for doing. The Greeks had a second word for time, *kairos,* which is eternal time. They knew that there were moments when we break out of ordinary reality and receive an epiphany or recognize an opportunity. Dreams emerge from kairos moments of epiphany, but then they have to be lived out in chronological time with people who may be resistant to change.

While this lesson has focused on gaining the ability to name the negative aspects of Cronus and Zeus so that they do not swallow you, upcoming lessons will explore how the positive sides of this power archetype can help you develop the ego strength necessary to utilize archetypes, including this one, as allies to assist you, without them running your show. When integrated with the lessons of Persephone, the Zeus archetype also can help you experience kairos moments without being overwhelmed by them, and then do the long, hard work of making the visions they presented to you realities in linear time.

~~~~~~

APPLICATION EXERCISE:

Creating an Inventory of Internalized Expectations

In a journal, in your mind, or in a conversation with a friend, take an inventory of all the ways you may be swallowed by family expectations, those of the society around you, groups of which you are a part, or your own beliefs about what you need to have and be to be a worthwhile person. You can do this in list form, and at the end, summarize the list, starting with this trigger: "What I can see driving me and making me feel empty inside is . . . ."

Zeus Lesson Two:

Declaring Your Independence

R EADING ZEUS'S STORY AS A PARABLE, we can recognize
that the threat of being swallowed is a metaphor for being
engulfed by one's family's worldview, desires, and injunctions.
Zeus's exile in Crete takes place during his maturation years, when
he needs to separate himself and clarify his own dreams, beliefs, and
desired legacy. His living alone in a cave relates to the part of us that
is ours alone, where we hatch our dreams and aspirations. This lesson
focuses on how to discard damaging and limiting narratives that are in
the way of your knowing what you want to do and achieving it.

A modern series of novels, Jean M. Auel's Earth's Children, set in
the Cro-Magnon period, depicts this process. The heroine is Ayla, the
daughter of very early *Homo sapiens,* who is the only survivor from her
group. She is taken in by the Clan, Neanderthals who do not speak,
cry, or laugh but do have sign language. Their social roles are divided
rigidly by gender. Females are forbidden to even touch weapons used
for hunting and must be available for sex with any male Clan member who makes a sign signaling his desire. Ayla has capacities beyond
those of the Clan members but also some liabilities. For one, she is not
physically as strong as many of them.

She also has an urge to use the gifts she has, and, watching hunters, realizes that this is one way she could make a bigger contribution to the group, so she teaches herself to hunt and becomes quite proficient at it. She bears a child after Broud, the son of the Clan leader, forces sex on her out of hatred, likely fueled by envy at her range of abilities and a desire to humble and humiliate her. The Clan members regard her hybrid son as having birth defects and want to kill him, but she works to save him. They also discover that she is hunting, thereby breaking a major Clan taboo. The Clan ultimately banishes her into exile, where she survives with her son through her hunting skills, eventually meeting a traveling *Homo sapiens* male who becomes her mate for life. Her challenge then is to unlearn many of the Clan ways in order to join human society and not be perceived as too odd. Fortunately, the Clan medicine woman had trained her in healing arts, so her gift is in demand, even though she uses skills that humans are unaccustomed to, eventually complementing Clan methods with human ones.

At some point, most adolescents today see older people as their generation's Neanderthals and strive to figure out how, in the face of these old-fashioned, "oh, so yesterday" people, to realize the full range of their abilities. Their strength during this period comes from bonding with others of their age who are facing similar challenges. In fact, we can experience this dilemma at any time of life when we feel as if we are carrying the gifts the future needs into unfriendly territory that seems to us hopelessly lost in the past—hence the lasting popularity of the Earth's Children series.

In *The Hero with a Thousand Faces*, Joseph Campbell defined Holdfast the dragon as the power of the past to imprison the present, and he went on to stress how the hero's or heroine's job—meaning the task for you and me—is to bring life to a dying culture. Doing so requires the narrative intelligence to know what stories or levels of stories have become anachronistic. James Hillman, in his bestselling book *The Soul's Code*, describes how each of us is born with the seed of potential, just as the acorn holds the potential of a giant oak. Out of this seed comes your optimal life script.

However, sometimes it is difficult to find that seed among all the weeds that obscure it. Weeds are not necessarily bad plants; they are just plants that are not supposed to be growing in a particular place. In our psyche, the weeds are internalized ideas from family, friends, and the culture that do not fit us. Weeds also are the inner self-critical voice most of us experience that compares ourselves to others, whether or not being more like them is right for us or would bring us greater happiness. Our task is to find the seeds of our potential, cultivate them so that they bloom, and weed out the unwanted plants that take away light and nourishment.

The need to discover our specialness is innate in our species. As children, we want our parents and others to notice and praise us as we gain new skills and abilities (e.g., "Look, Ma, no hands!"), and we rely on others to mirror our strengths back to us. If we, like Zeus, do not have access to this kind of parental reinforcement, we may have an even stronger need in adulthood to achieve spectacularly visible things than those whose parents showed joy in their accomplishments; failing that, we may become either extraordinarily needy or troublesome. In any case, we will want to be center stage, so best to do this consciously.

In the process of differentiating, it is normal to become aware of the places where we do not agree with our parents and other authority figures, and where their ideas for us may not be congruent with our own. If our parents want us to be true to ourselves, forging our own path is unlikely to create a fissure in the relationship; but even so, we might hesitate to share these differences honestly if we think they might make them unhappy. If, with an overabundance of Demeter care, our parents have trouble letting us solve our own problems, we may start believing that we are incapable of doing so. Since it is every generation's role to bring new life to the culture, inevitably we need to go our own way, even when—or especially when—we feel scared of doing so. That is why we often must educate parents and others to understand that we consciously are making a choice different than the one they would make for us—and for good reason.

The young, along with historically underrepresented people of any age, share the archetypal process of needing to work to have their gifts be recognized. A culture with norms based on a privileged class, however, pressures us to prove ourselves by the standards of the prevailing group in power—which, in the case of the United States, still is largely older, white, male, heterosexual, and so on. Yet as dated and sometimes prejudiced as those standards may be, it is important to respect and absorb the knowledge possessed by those with more experience in running things.

Many religious laws provide guidance in understanding the logical consequences of this. For example, the Ten Commandments (from the Torah and Christian Old Testament) begin with an injunction to honor our fathers and mothers that our days might be long in the land. This is not just a moral statement; it is one of logical consequences, implying that you and your generation need the wisdom and experience of the elderly to survive, and it also holds up the implied threat that if you do not care for your parents, they might not be around to take care of you. We can extend this to the practical current wisdom of honoring your boss, the CEO, and your mentors that your days might be long in the company—at least until you are ready to leave.

We all stand on the shoulders of our ancestors, so differentiating should not mean that we fail to honor what has gotten us to where we are. However, a robot-like, death-in-life feeling of no energy or zest for living may be a signal that some old idea or allegiance is holding you back, or that there is a wound you need to confront and heal. Differentiating yourself from your parents, and becoming your own person, is an essential task of turning into an adult. You do this by questioning assumptions others take for granted; ask yourself, "Money for what? How much do I need to fulfill my calling and enjoy my life? Power for what? How much and what kind of power do I need to fulfill my purpose? Status for what? How important is status to my life? Who do I want to think well of me? Well-being for what? What qualities do I need to embody and what experiences do I need in order to develop the capacities required to live my optimal life?"

It is not just what others demand of us that swallows us at every age; we are always emotionally inhaling what is said and done around us. Our psyches are porous. Absorbing societal attitudes is a part of the human condition. The human desire to belong is so strong, according to psychologist Abraham Maslow's hierarchy of needs theory, that it comes right after the need for safety and security.[1] Yet our ability to think for ourselves and to find our authenticity requires us to become aware of our unconscious programming and sometimes push back against the norms of people we love and care about. Every group or organization of which we become a part has written and unwritten rules that are elements of an implicit storyline that may or may not be right for us. The secret to healthy belonging is to be able to be fully part of a group yet still be true to our nature and values. This means that we need to say no to the group when it promotes unhealthy behaviors and attitudes. Otherwise, we fall prey to what Jung called the participation mystique, where our authenticity is swallowed by the views and actions of the group. Often, learning to speak our separate truths can save work teams from ineffective groupthink where people pretend to agree in order to fit in. When all views are not heard, groups can make very bad decisions.

As we grow, we constantly are being exposed to cultural, political, and family narratives. Until we gain the narrative intelligence to recognize what stories we are taking in, telling, and living, they run us, and most of the time, we do not even notice. This process continues over the course of our lifetimes, since we always are being socialized by the people around us and by the larger culture. When we develop our critical faculties, we begin to be able to question the advice of parents, teachers, ministers, friends, and bosses, and even our own thinking. And because we do this based on the level of our understanding of ourselves and of the world that we possess at any given time, we sometimes may get it wrong, but then, ideally, we learn from our mistakes.

Most likely you are well aware of the impact of Madison Avenue on your consciousness, and you may be working to be a conscious consumer. It is important to recognize that in selling you things, marketers also are influencing cultural ideas about what is necessary to be successful. We can be exhausted from trying to live up to these pressures to be the perfect parent, in physically great shape, attractive at any age, professionally successful, a socially conscious volunteer, up to speed on major social and political issues, and so on.

The ideas and stories we have swallowed also can give us a stomachache or otherwise make us ill. Nurse and medical intuitive Christel Nani, in *Sacred Choices: Thinking Outside the Tribe to Heal Your Spirit*, reveals how she found that clinging to what she calls "tribal beliefs" that do not serve one's authentic thriving actually causes illness.[2] She recounts inspiring cases where letting go of such a belief and being truer to themselves helped people to get well. I cannot tell you how many individuals I've known who got sick when they were stuck with a job, a marriage, or another situation that stifled their creative self and then experienced a miraculous recovery when they did what they really wanted to do.

One example of a damaging tribal belief Nani illustrates is the dictum that a good woman takes care of others—especially her family—before herself, which we can recognize as a Demeter trap. This common tribal belief can lead directly to less than optimal health for a woman until she reframes it to care for herself first. A similar idea that a man or woman unconsciously controlled by his or her Cronus parts might have is "I always need to stay in control; otherwise, I or others may get hurt," leading to chronic stress and a similar breakdown of health.

Whatever story holds you back, your task is to shift to one more empowering, though that is not always easy to do. The most challenging form of narrative mindfulness requires being able to notice and critique the stories you are quietly telling yourself in your innermost thoughts.

Even families that look ideal from the outside can wound their children inadvertently, and most of us have a psychological scrape or two to deal with. I know that many little girls want to be princesses, but the reality of being royalty or rich is not always what we imagine it to be. The Academy Award–winning movie *The King's Speech* is based on the true story of King George VI of England. Before he becomes king, George (known as Bertie) does not appear to be a good candidate for the throne. For one thing, he has a severe stutter. (Interestingly, this parallels an archetypal biblical story in which Moses, brought up in Pharaoh's house, asked God not to send him to rescue the Israelites because he had a stammer, but God chose him anyway because of Moses's kindness to the sheep he herded.)

Bertie, who had been second in line, is thrust into the kingship a few years before the Second World War breaks out when his older brother relinquishes the throne to marry an American divorcée. Imagine that you cannot even make a statement without stuttering, and your role is to inspire a whole people through public addresses. Bertie is in despair about his ability to do so, but his wife, Elizabeth (later much loved as the queen mother), finds an unconventional but effective speech coach to work with him.

Operating much like a psychotherapist, the coach, Lionel Logue, takes Bertie back to the pain of his childhood—when he was habitually mistreated by a governess, ridiculed by his father and brother, and in agony because his legs were in braces night and day to straighten them out. If that was not enough, a younger brother, whom he loved and who died young, was kept from the public eye because he was epileptic. The pain in the royal family came from their desire to maintain their image of how royalty should be perceived, a situation not unlike that of anyone coming from a family that felt the need to appear a certain way.

Reliving his suffering, Bertie begins to understand why he lost his "voice," since his very selfhood had been suppressed to stay true to his family's script for him. As he recognizes the hurt and outrage felt by his child self, he begins to find himself for the first time. Gaining empathy for himself opens his capacity to empathize with the fear and pain of British citizens as they endure the dangers and privations of war. His dignity and care show in his voice and words, so much so that his speeches are still remembered, even though he always struggled to speak without a stammer.

Lionel's intervention also helps Bertie reclaim his royal heritage. Lionel angers him by sitting down casually on the throne that Bertie will occupy after his coronation. When Bertie tells him not to sit there, Lionel asks him rather insolently why he should listen to him. Bertie lashes out, shouting that he is to be listened to because he has a voice. In this powerful breakthrough moment, he speaks with the voice of a king, not that of a stammering child.[3]

So, too, with us. Whatever the traumas of our early lives—whether great or small—the process of healing through facing them awakens empathy for others and makes way for a renewed sense of purpose. Often, that purpose combines our native strengths with a legacy from our family. I remember a woman who attended a training I conducted who came from a very dysfunctional family, with a mother who was mentally ill. Claire was quite wounded psychologically, yet she walked like a queen. Whatever their defects, and they were many, this family was very prestigious, and from them, Claire had gained an intrinsic sense of entitlement that helped her trust that whatever she wanted to do in the world she could achieve. At the same time, her wounding motivated her to follow a shamanistic path, and thus she became a wounded healer who could help heal others.

Claiming our independence makes us the monarchs of our own lives, but this accomplishment entails duties and responsibilities. It is not all about doing what we feel like doing but rather about doing what we know we must do, because it is our purpose to do it. The Demeter archetype helps with the task of finding your purpose, since this journey begins with discovering whom and what you care about. The Zeus archetype helps you build on this discovery and locate your strengths, so that you can be a contributing and responsible member of society.

Figuring out what you want sounds simple but actually can be quite difficult. The challenge is to know yourself well enough to recognize when well-meant advice is not right for you. Many of us realize in retrospect that we spent a great deal of time studying what someone else wanted us to study or engaging in activities just to please our teachers or our friends—however stressful we found them, since they were a bad match with our innate abilities. The stakes get even higher when a decision involves committing to a life partner or discerning what might be our vocational calling. (My high school guidance counselor enrolled me in the clerical administrative track until my English teacher intervened and told me I should go to college, saving me from a profession at which I would have been horrible—but hey, as a writer, it is a good thing that I learned to type quickly.)

Personal challenges in declaring independence mirror societal ones. Dreams take time to be realized, whether for you or me or for entire countries. America's founding promise is an example. The country's founders declared, in the 1776 Declaration of Independence,

> We hold these truths to be self-evident, that all men are created equal, that they are endowed by their Creator with certain unalienable Rights, that among these are Life, Liberty and the Pursuit of Happiness.

Before the signing of the Declaration, Benjamin Franklin warned his fellow delegates to the Second Continental Congress, "We must all hang together, or assuredly we shall all hang separately." Yet those brave men (and the women who supported them) pledged "our Lives, our Fortunes and our sacred Honor" to make the promise a reality, and many of us today are still working to fulfill it—happily, in most cases, without such dire consequences hanging over our heads. When we are living someone else's story, we may feel like a colony and also be fearful of declaring our independence. Even though we are not likely to be killed as a result, breaking away may require a commensurate amount of courage.

When I think of others throughout history who have stood up to make a difference, Martin Luther comes immediately to mind. He followed his conscience and incited the Protestant Revolution. I love that he is widely credited with declaring, as he broke the rules to be true to his beliefs, "Sin bravely so that grace may abound"—even though he likely never actually said those exact words. I also am inspired by his namesake, Martin Luther King Jr., whose ability to share his dream of a country where people are judged by the quality of their character, not the color of their skin, galvanized widespread support for the civil rights movement.

Such audacity has worked in business, too, evoking mythic overtones. You might have heard about or seen the 1984 ad that put Apple on the map. In this ad, which was a takeoff of George Orwell's novel *1984,* a forbidding-looking man on a screen in front of an auditorium (representing IBM as Big Brother) is telling robot-like people in the audience what to do. An athletic young woman (the heroine) runs in and throws a sledgehammer, smashing the screen and freeing the people to be alive and fully human. The voice-over says that this is why 1984 will not be like Orwell's *1984*. Then the Apple logo, with the usual bite out of it, appears, making a mythic connection with the apple in the Garden of Eden. This logo, with the tagline "Think Different," implicitly calls up the theological doctrine of the "fortunate fall," which challenges us to rebel against the dictates set down by authorities (with the highest being God) and instead to exercise free will governed by conscience, with its knowledge of good and evil. In evolutionary terms, the fortunate fall can be reframed as humankind becoming conscious, not defined just by instinct. But of course, at the literal level in the world of commerce, in the ad Apple is comparing its user-friendly products to IBM's, which are depicted as forcing users into Big Brother predetermined patterns.

Building on this lesson, which has encouraged you to declare your independence, the next section examines how you can develop the strengths and capacities that enable you to succeed in your own unique way.

APPLICATION EXERCISE:
Declaring My Independence

As a life practice to use when the expectations of others—family, friends, coworkers—undercut your authenticity, first define how they do not fit for you. Then declare your independence, perhaps pledging your life, fortune, and sacred honor that you will be true to yourself: "I, (your name), hold these truths to be self-evident, that *all* people are created equal, and since *I* am a person, *I* have an unalienable right to my own life, as well as the liberty to make my own choices and to pursue happiness in my own way."

Zeus Lesson Three:

Unleashing Your Passion, Focusing Your Actions

T HIS LESSON EXAMINES how you can unleash the power of your genuine passion and tap into your intelligence and good sense to focus your energies in a positive direction while guarding against acting impulsively when a wiser move would be to shift to a different story or story level. But first, to tap into Zeus as a mythic model, let's consider some additional details of Zeus's life as a young adult. When Zeus goes to Crete, he is an exile, a powerless child. When he returns to Olympus, he is so powerful that he often is depicted as, or with, a bull. In ancient times, the bull connoted positive momentum, and it does now as well: a "bull market" means stock prices are on the upswing. A bull also embodies the fertility and fierceness of the archetypal fertility king described by Sir James George Frazer in *The Golden Bough*. In this mode, Zeus commits to overthrowing Cronus and becoming the chief god.

Many people starting out, or starting over, feel very much like Zeus in his cave, exiled from the center of power, especially if they are back living with their parents because work is hard to find, are stuck in a job that is not what they would have chosen, or have just lost a job, their money, or their means of support. Yet such times present an opportunity to clarify goals, develop physical strength through exercise, and sharpen mental focus through using affirmations to reinforce an attitude of "I can, and I will!" It also is a time for making contacts, identifying options, and assessing their viability.

Even when we are very engaged in the world, we also need a place to retreat to, the cave where we can regain clarity and restore our energy. I believe the story that helped Zeus do this might help you too. The story of the Minotaur was a primary Minoan myth that anyone of that time would know. It began with King Minos and Queen Pasiphae of Crete asking Poseidon (the god of the sea) for a white bull as a sign to show the people that the gods blessed their reign. However, when they received the bull, they failed to thank him. Poseidon enlisted Aphrodite to help punish the royal couple for their lack of gratitude and humility, and they caused Queen Pasiphae to fall in lust with the bull. Acting on this powerful attraction led the queen to give birth to a child, a Minotaur, who had a man's body and a bull's head. He was so fierce, ruthless, and dangerous that a labyrinth (more like a maze) was designed to keep him imprisoned underground. When the Minotaur demanded that people be sacrificed for him, Theseus, a warrior hero and king of Athens, killed him (with the help of Ariadne, who later married Dionysus).

Zeus likely came to understand that the problem with the Minotaur was that, with his human body and bullish head, the wrong part of him was making decisions. The inner bull in us, experienced as the power in our bellies, or as unexamined gut impulses, is surrounded by organs and a maze-like digestive system that parallels the image of the bull in the labyrinth in the story. Knowing this can provide a metaphor to remind us to assimilate and integrate animalistic impulses before acting and then to check with what our brains would advise. We can imagine such an inner bull as the energy source for our superpowers, but not as their guidance system.

But what happens if we keep the lid on our superpowers entirely? This would be like getting stuck being the alter ego of any modern superhero. This dilemma is depicted in a humorous way in the animated movie *The Incredibles*, in which society has turned against superheroes and wants them to act normal. The first time we see Mr. Incredible—a man with amazing strength—he is sitting at a desk way too small for his girth, processing insurance claims and feeling despondent. He has been reduced to a cog in the wheel of a vast insurance company that denies little old ladies their due. As the movie continues, we learn that this is to the detriment of him as an individual, the society at large, and the other members of his family, all of whom also are miserable as a result of repressing their superpowers. What frees them to reveal their powers and good spirits is the need to protect their city from total destruction.

The Zeus part of you wants you to release your own superpowers—meaning your gifts and wisdom—to address the problems you see in the world or to realize the dream you have for what life could be like, for you and for others. But for some, that dream hovers over them as an idea, lacking the oomph that's needed for them to get up and get going to achieve it. Years ago, I collaborated with a psychologist and Iranian immigrant, Chris Saade, who would dramatically—with a booming voice arising forcefully from his belly—enjoin people to find what they "de-si-er," as he would say it, drawing the word out into three long syllables. With his charming accent, coupled with the intensity of how he spoke and moved, he modeled what it means to know what you want from a very core and primal place. His example helped us let go of ambivalence ("Like, I sort of want to . . . whatever") and fully claim the power of our desires. Tracking when you feel energized can help you identify your strengths and life purpose.

Experts like Mihaly Csikszentmihalyi (*Flow: The Psychology of Optimal Experience*) have researched experiences of thriving, enabling us to better recognize moments of what he calls "flow." For example, in a flow state, you gain energy from what you are doing and lose track of time. You feel completely "in the experience."

So what do you desire at this moment in your life? You likely already know a good bit about it. For example, you might recall how sometimes you sit, feeling lethargic as you think about something you should do. You tell yourself to get up, but you don't move, or if you do, only grudgingly. But if an opportunity comes up to do something you really *want* to do, you find yourself leaping out of the chair and getting going. You also may experience a similar spiking or sinking of your energy level in your gut in response to signs for the "right path" and "wrong path."

If your Zeus is paired with Demeter, commitment and determination provide energy to persist in making a difference even when it is hard to do. A high-level executive in the U.S. Department of Health and Human Services shared with me how she kept her energy and that of others in her office up by always returning to what they all cared deeply about: aiding children in poverty (clearly informed by the Demeter archetype). I was familiar enough with her work to know that this Demeter calling was buttressed by very fine Zeus-like administrative abilities. She was able to keep her whole team focused and effective in helping children even when the administration of that time and its political appointees were setting goals and policies that sounded good to their constituents but were known to be ineffective or even irrelevant.

More often than not, responding to a genuine calling gives us a shot of energy, whether subtle or intense. The trick is to distinguish this jolt from the hit we get from anxiety, greed, lust, or fear. Danger signs stimulate the amygdalae in our brains; therefore, threats get our attention. Our brains are wired this way so that, in the wild, we would gear up to run from the tiger, or now, in urban environments, we can get out of the way of a train or bus careening toward us. We can be filled with excitement in the belief that we have found our true love—but actually, the energy burst may be adrenaline activated by some deeper part of us warning of danger: "This guy or gal will turn out to replicate a past difficult relationship with my mother or father or some other important formative figure." Or "This new work situation will replicate the trauma in the one I just left." Conversely, you may feel anxious before giving a talk, yet adrenaline can be helpful if you tell yourself that this is a rush of anticipation to mobilize your awareness by saying, "Yes, I can and will do a great job! The adrenaline is getting me ready," rather than thinking, "I'm so anxious, I'll mess up and humiliate myself."

Much of what we see in the media is designed to increase ratings, not to help us understand the complexity of today's world. Being continually pummeled with news of dangers on all sides stimulates the bull in us and increases our sense of anxiety, outrage, and eventually, cynicism. Wise heroines defuse their unproductive bull energy by choosing news outlets that provide in-depth journalism that activates the cerebral cortex, analyzes situations, and evaluates solutions that could support good public-policy decisions. When the steady onslaught of bad news keeps coming at us in a way designed to entertain and rile us up, it is good to remember to breathe deeply, calm down, and take action on the issues that call to us, even if only by writing to our representatives in Congress.

What most upsets us also can become a clue to our mission. Our passion for that mission then serves as the fire in our bellies, but we also need to stop, cool down, and think if we want to act wisely to achieve our goals. The point is not to get tied up with negativity but to shift from recognizing a problem to figuring out what we can do to be part of the solution.

The story of the Minotaur is a cautionary tale of the need to bridle the fierce and lustful bull potential we all have inside us in order to avoid impulsive and counterproductive actions. It is important that we ride the bull, and that it not ride us, but it also is critical that we not repress it completely.[1] A bridled inner bull provides the fuel for our journey, but our minds still need to determine where we go.

Gaining the ability to act on his desire to overthrow his father and become the chief god of the gods, Zeus also is able to focus his energies well enough to win his war against the Titans, who are the older generation, aligned with Cronus. But the result of the war is a kingdom in disarray. It is clear that he still has much to learn, so he strategically cultivates wise counselors who can assist with his ongoing learning. From them, he gains the skills and perspectives needed to rule successfully, beginning with Metis, who helped him win the war, and then Athena, who helps him rule Athens as a city-state that becomes the cradle of democracy.

Although he has overthrown Cronus on the outside, on the inside, he still is acting like him. For instance, Zeus swallows the goddess Metis, the advisor who helped him achieve victory over the Titans (and who in fear had turned herself into a fly to get away from him). He also sentences Prometheus to harsh eternal torment for stealing fire from the gods and giving it to humankind, thus appearing to be a tyrant and losing some important political capital (although Hercules rescues Prometheus after Zeus cools down). His emerging challenge is to become a more benevolent ruler and in this way cement loyalty out of respect rather than fear. In this, he is helped by Athena, his daughter by Metis.

A danger for advisors is that people seeking their help will expect them to solve problems for them and will fire them if they do not. A literary example of what should happen instead can be found in *The Wizard of Oz*, in which Dorothy expects her sidekicks to lead her to the great Wizard, whom she believes will solve the problem of getting her home. Each of these characters symbolizes the deficiencies she believes she has and the real strengths underlying them. The Cowardly Lion saves the day; the Tin Man, who lacks a heart, is always weeping with touching empathy; and the Scarecrow, who lacks a brain, makes the plans. When they find the Wizard, he cannot solve her problems or theirs; he can only validate these various strengths. Dorothy finally learns from the Good Witch (of the South in the novel, and of the North in the movie) that what mattered was that *she* gained the lessons she needed on the journey, so now she can tap her heels together and go home, a new and more competent and grateful person. Similarly, good personal and professional coaches today do not solve our problems; they help us learn to do so ourselves.

Athena is the goddess of wisdom, after whom Athens was named. When Zeus swallowed Metis, he also swallowed the fetus of Athena. He then incubated her in his head, which metaphorically suggests that she is part of him, as well as a separate goddess who gives him the worst headache he ever had. When she emerges from his head, she already is grown and in full armor. Understanding the strategies needed to protect Athens from invaders, she combines clarity of intent with perseverance, courage, and the cleverness to advise Zeus about how best to govern the city-state. She is the goddess of vehicles for getting places, such as the chariot and the ship, and is credited with inventing the bridle, which allows for the taming of horses so they can be ridden, and symbolically the bridling of the animal part (bull, elk, horse) of the Zeus archetype. Thus, she can help Zeus use his intelligence to curb the excesses related to his hypersexuality and aggressiveness, so that he can ably manage the affairs of state.

Many people get riled up about things they wish they could change or have fantasies about what they want in life but have no structure, and no support, to help them achieve their goals. Thus, they do not actually move forward to solve problems and live the life they dream of. The wisdom of Athena, useful to Zeus and to us today, includes the narrative intelligence to visualize what we want in the kingdom of our lives, how we might move toward it, and how we might get the support we need to persevere in realizing our visions.

Years ago, a friend and I committed to sharing our intent for the upcoming year every January and to checking in monthly to see how we were doing. Our goal was to become conscious of our desires and to act upon them. We conceived of this as different from New Year's resolutions, which generally are things people think they *should* do—lose weight, begin an exercise program, be more organized, and so on—and can accomplish through will and discipline.

Because our purpose was to fuel this kind of success by the power of authentic desire rather than will, we allowed ourselves to change our intentions. Often, our lack of energy for actually accomplishing the vision tipped us off that what we had said was not really what we wanted to do. We then would refine those intentions to bring them in line with our desires, so that our actions could naturally support them. Sharing such intentions with one another helped us to anticipate potential problems, and if unexpected ones occurred, to strategize about how these might be handled. Most of all, it worked to focus our energies, so that we did not get too distracted by transient desires or the pull of immediate crises. We have been doing this for years, with the result that we have become professionally more successful and personally more fulfilled.

Support around clear intentions is remarkably effective. People with severe illnesses who belong to support groups tend to have better outcomes than those who don't, and twelve-step groups help people free themselves from various addictions. A project in San Quentin State Prison (Guiding Rage into Power) has shown that group sharing can help even lifers find peace with what they have done so that they form close bonds with fellow inmates and transform their experience of incarceration. In that process, they stop blaming others for their own misdeeds, instead taking responsibility, feeling remorse, and then committing to becoming a different and better kind of person.

Real transformation happens to the Zeus within any of us when our focus shifts from what needs to change in others to what needs to change in ourselves. When we become the monarchs of our lives, it is easy to think of life in hierarchical terms and come down hard on anyone who seems to challenge our decisions or ideas. Less-developed monarchs, therefore, often see people who challenge them as enemies to be quieted or squashed. Zeus's first response to Demeter's sit-down strike was to assume that he was right and she was wrong and the solution was for her to shape up. A potential failing of the Zeus part of us is the tendency to project inadequacy and even blame onto others, while seeing oneself and one's group as superior, even blameless. It is the way we shore up our egos and blind ourselves to what we do not want to see in ourselves.

Lawyer and author Louis Nizer is famous for saying that "When a man points a finger at someone else, he should remember that four of his fingers are pointing at himself." When you find yourself judging another person, it is useful to examine whether he or she may be mirroring something in you that you should pay attention to. In my life, I've noticed that I am more likely to reflect on others' misdeeds when I'm feeling guilty, ashamed, or insecure. I also have learned to stop and think when I catch myself judging someone who is violating a value of mine. Usually, this is a wake-up call for *me* to either live more fully by my values or to know that I cannot countenance what I'm seeing. The latter sometimes requires a Demeter-like exit from the situation, but it also can start with letting go of my judgment of the person or people in question, realizing that I'm not perfect and I do not know what it is like to walk in their shoes. It then is easier to come up with a reasonable and respectful way to handle the values conflict.

Zeus eventually recognizes that he can nurse his anger at Demeter all he wants, but it will not resolve the situation. The famine abates when *he* changes. The greats among us have demonstrated an ability to end the cycle of atrocity and revenge by determining that it stops with them, and they have done this even in times and places where they have been horribly oppressed and abused. Nelson Mandela, after being incarcerated for 26 years, declined to keep the cycle of revenge alive. As he told Hillary Clinton, "As I walked out the door toward the gate that would lead to my freedom, I knew if I didn't leave my bitterness and hatred behind, I'd still be in prison."[2] This decision enabled him to become president of South Africa and oversee the official ending of apartheid. What might you or I do if we let go of all our justified resentment so that we could focus more completely on being the change we want to see? How might we be even more effective, then, in solving the real problems that caused our suffering and that of others?

Having explored how we might best claim the power to rule our lives and then learn to direct them more strategically, in the next section we will examine how we can continue to learn more as we go, and thus refine our sense of direction for the next steps we need to take.

~~~~~~

APPLICATION EXERCISE:
Becoming My Best Self

Sit comfortably, breathing in and out slowly and deeply, and then imagine that you have "made it." You have achieved your ambitions. To make this really tangible, visualize a day five years from now when you have acted on your clear intent to live the perfect life for you: picture where you are, who else is around, what you are doing, and how full of joy and energy you feel. Then envisage a story that gets you to this happy outcome, considering the following questions: What challenges do you think you will face? What skills and abilities will you need to utilize to meet these challenges successfully? What advisors, coaches, or role models will you need and how can you receive what they offer in a way that awakens their abilities in you? What temptations do you face that make you want to be rescued or to find someone else to blame, and how can you resist them? What trauma or grievance do you need to let go of? What practices of personal excellence will help you get to where you want to go? (This is the kind of fantasy that can live in your mind over time as you ruminate and achieve greater clarity, so do not worry if it seems difficult to complete this exercise quickly.)

Zeus Lesson Four:

Regrouping and Rethinking as You Know More

THIS LESSON EXPLORES HOW, when challenges grow in intensity, it is normal to be stumped at first and to find yourself replicating old behaviors in such a way that you only worsen your situation. Just when we have mastered living one story, it very typically stops working for us, leaving us with the need to regroup, rethink, and live a new one.

For Zeus, winning the war with the Titans and displacing his father was not the end of his strife. Various myths tell us that Zeus's reign was undercut repeatedly by the continuing anger of the Titans, who never truly accepted their defeat, especially when Zeus was still a novice ruler. It was they who created Typhon, the most awful monster ever. Some myths said this monster was born of Gaia, the primal Earth Mother, perhaps to save the world from human exploitation. Some said it was Hera, Zeus's wife, in her immensely powerful Minoan form, who created it to punish Zeus both for defeating the Titans and for his infidelities.

Typhon, whose name is the basis of the word typhoon, was a storm god who had the power to stir up the weather and create chaotic conditions. In their first encounter, Typhon defeats Zeus, removing or destroying the tendons in his hands and feet, leaving him powerless. But Hermes, the god of commerce and communication, replaces the tendons and saves Zeus. Healed, Zeus thinks quickly and calls down lightning and thunder, organizing Typhon's chaotic storm and focusing it toward Typhon's head, which then catches on fire. The earth opens and Typhon is trapped under Mount Etna, where he becomes a volcano.

Through meeting this challenge, Zeus lets go of living a war story to claim his power as a sky god and ruler of Olympus, as any of us do when we move from fighting to get in the position we desire to actually doing what it requires. The job of the king, queen, or boss is to bring everyone and everything to a place where they can be most useful, or failing that, where they can do the least harm, as happens with Typhon. The fact that it was Hermes, the god of communication, that healed Zeus suggests that Zeus learned to trust the power of communication to inspire others, rather than turning to violence. In this way, he shifted the core story that determined the lens through which he saw the world. Communication and negotiation are the skills of a ruler, not a warrior.

After helping to defeat the Titans, Zeus's two older brothers, Poseidon and Hades, could have challenged Zeus's reign, but he emerged as the chief god through his wisdom as a negotiator and his understanding of power politics. Instead of going to war, the brothers decided civilly to draw lots. Zeus won the right to be the ruler of the earth and sky, while Poseidon contented himself with governing the seas and Hades the Underworld. Zeus's older sisters, Hera, Demeter, and Hestia, presided over traditional female support roles: wife, mother, and homemaker, respectively. In resolving these issues, Zeus used negotiation and strategy to clarify roles and responsibilities in a transactional, instrumental leadership way.

Thinking of Typhon as the origin of typhoons might remind us of times when our lives seemed so busy, chaotic, or difficult that we felt overwhelmed and could not keep up. Such occasions often come when we let go of the superwoman narrative and allow another story to well up within us. Or we may be stopped in our tracks as we deal with career or life setbacks and recognize that what we learned from the experience made us grow in ways we might never have otherwise.

The *power to* achieve goals and create new, more desirable outcomes often requires us initially to fight for rights and show that we are strong enough not to be bullied—that is, to live out a war-story plot. This can be appropriate when we lack access to what we want, and need to struggle to get it, or when we go to the rescue of others who are being mistreated and who need our support, and in the process demonstrate our strength. Once we have succeeded, we can shift from living a war story to a ruler story, finding a way to heal the divide and bring people back together.

This is true for countries as well as individuals. Our current context (in the United States and Europe) began with World War II, when the United States and its allies demonstrated the strength and resolve to defeat Germany and Japan. After the war, the Marshall Plan rebuilt Western Europe, benefiting our wartime enemies as well as our allies (although not the Soviet Bloc, which rejected the aid). This wise and generous act stopped the cycle of revenge in Europe. Having demonstrated military strength and moral generosity, the United States became a superpower. But then, inevitably, new challenges emerged.

Following World War II, the Soviet Union, our former ally, became our major enemy, with the world choosing sides. One strategy for maintaining peace became the nuclear-arms race, which made another world war unthinkable. Then, after the Cold War abated, there was a brief reprieve until 9/11, when terrorism became the world's greatest problem. With terrorism came the increased complexity of tracking even potential threats, since terrorists don't observe national borders and can blend well into civilian populations. In addition, the arms race that was part of the Cold War, and may have helped end it, created the threat of nuclear proliferation and weapons of mass destruction getting into the hands of terrorists.

When we don't know how to face a new and challenging reality, most of us retreat to what we do know until we can determine how to respond. In my consulting practice, I've observed how many smart leaders understand that their organizations need to innovate more quickly but cannot figure out how to win the internal political battle to change structures that inhibit creativity and keep things slowly plodding along. This is another reason many of us are so overwhelmed and exhausted. We live in multiple systems that, though new stories are needed, continue to live out the old scripts, doing what they have always done, only longer and harder. The result is a little like running on a treadmill. Everyone expends a lot of energy, but they don't get anywhere.

The Zeus part of us seems to be attached to stories where we are the good guys and are victimized by the bad guys, so we focus on defeating them rather than on achieving what we truly want, which is especially the case when the real problems defy easy answers. When media pundits and entire political parties do not yet know what to do, they fight each other, creating the illusion that they are accomplishing something. This tendency can lead to an inability to track real dangers—one reason that 9/11 and the Great Recession of 2007–2009 came as surprises. Even though climate change is no surprise, the political battles surrounding it mean that, at least according to a 2014 United Nations report, it likely will not be addressed until it is too late.[1] Similarly, businesses can be so focused on besting the immediate competition or on internal turf battles that they don't even notice an invention in the works that will make their whole field obsolete or leave it fighting for its life. For example, manufacturers of typewriters and business machines continued to battle each other for market share even as computers were coming into more common use. Later, IBM, which led the way in the commercialization of computing (the company adopted the initials to escape from the narrow connotation of its full name, the International Business Machines Corporation), dismissed the notion of the personal computer, believing that mainframes would continue to dominate the market. Of course, part of the genius of Apple, at this writing the most valuable company in the world, and of its cofounder, the late Steve Jobs, is its constant focus on what its *next* innovative product will be, even as it continues to improve its current product line.

On the individual level, many people keep holding on to resentments against parents or former spouses, rather than moving toward the life they want to live. Today, members of the millennial generation often complain about the unreasonably slow and cumbersome procedures kept in place by older managers, while these managers criticize younger workers' ignorance of how to do things the "right" way. In truth, if either group stopped to think, they would realize that they need each other. The old guard needs the creativity, technological ease, and energy of the younger generation. And the younger workers need those who are more experienced to mentor them. If they put their minds together, their organization could optimize its success.

Similarly, sometimes we just have to fight for our rights in the short term, but this rarely solves issues in the long run. Liberation movements have been successful in changing many attitudes and laws, but racist, sexist, homophobic, and other biased attitudes persist and are given new energy as people respond fearfully to the changes. Individually, we might win a battle with our boss and later find we have moved into a cold war. Long-range success requires a firm stand followed by true active listening to win the peace. This is why women need to live more of a ruler than a war story, since our goal typically is to work in partnership with men, not defeat or be in perpetual conflict with them. The same goes for class divisions. A successful strategy is one that creates win-win solutions that consider the interests of all classes.

However, the goal is not just to win or find a compromise. Rather, it is to discover a third story that moves everyone involved beyond a "my story versus your story," either/or stalemate. When Zeus learned how angry and upset Demeter was, he could have listened and learned, reconnected her earlier with her daughter, and avoided a famine that almost wiped out the mortal population. Jung argued that even in our own internal lives, when we are trapped wanting two divergent things, we should hold the tension until an image or dream reveals a third option. In group situations, it may be even more important to hold this tension until another way is revealed.

In my leadership roles, I've also experienced an ongoing tension between the desire to make decisions by group consensus and accountability structures that hold the leader solely responsible for the results. Some places have tried holding the entire group responsible, but that is problematic too. If you ever had to do a group project at school where you did most of the work and the slackers in the group got the same grade, why this often fails to work will be obvious to you. A potential gift of the Zeus archetype is the ability to empower others and harvest their intelligence while taking the level of responsibility that is appropriate to one's own role, and requiring that others do so as well. It also is the facility to match observed talents to needs in any organization, so that each employee can make her or his optimal contribution to the whole.

An often ignored part of taking responsibility in a Zeus-archetype ruler story is to continue to develop your own capacities for self-assessment so that you recognize your own evolving right fit with the needs of your environment. Even the most accomplished among us experience setbacks or major failures at different points in our lives, which can spur us to develop new capacities and to redefine our goals. If we recognize that setbacks are normal, we can waste less time in self-doubt and open to new visions that take us closer and closer to our finest achievements. Remember that it was when Demeter, overwhelmed by grief and frustration, just sat down in her temple and waited that a resolution to her dilemma emerged.

When we are faced with greater and greater challenges, some of which throw us for a loop, it can feel as though a master sculptor is chipping away the parts of the rock that are not authentic to us to reveal our best selves, a process that can be painful but is easier when we do not fight it. A woman in one of my workshops shared her idea that the pieces that life chisels out of us shape the puzzle piece we are. From this, we recognize our partialness and how much we need others, so we go seeking complementary-shaped puzzle pieces that fit with ours. While initially we claim our power by differentiating from others, eventually we realize that our power grows as we find those who best complement our strengths with theirs. As we support their strengths, we benefit from them as they benefit from ours. Together, then, we can move mountains.

One of my personal heroes is Parker J. Palmer, a highly respected writer, lecturer, teacher, and activist. His life has been full of achievement, but it also has been peppered with times of depression, what he describes as just wanting to sit, which could be the metaphorical equivalent of Zeus losing the use of his hands and feet. While I often hear from very distinguished people that they have experienced failures that left them emotionally limping for a while, depth psychologists view such times as potentially generative, since they frequently lead to breakthroughs. It is generally when we fail to pause to regroup and rethink that our bodies, psyches, or circumstances step in to stop us.

Palmer has a strong commitment to living an authentic life nourished by his Quaker faith. His memoir, *Let Your Life Speak: Listening for the Voice of Vocation,* reveals, with candor and courage, some of the real-world difficulties of doing so. He was raised with the belief that anyone could grow up to be the president of the United States. (In the book, he confesses, rather sheepishly, that only later did he notice that "anyone," at that time, meant only white males.) Thus, he presumed that at some time in his life he would be president of something, and early on he demonstrated leadership abilities.

A fundamental value for Palmer is his commitment to developing community—a value that is key to his calling. Although he could have had a successful academic career (he has a Ph.D. from the University of California, Berkeley and had a job offer early in his adult life to be a professor at Georgetown University), he decided to go into community organizing, which seemed to be a better fit for his values. However, like many professionals today, he soon realized that what he thought he wanted was not the right fit for him. After discovering that he was a bit too sensitive for the rough-and-tumble world of community organizing, he did not know immediately what to do. But soon the energy to move returned when a new dream emerged: recognizing that he needed some time to learn about what living in a genuinely supportive and interdependent community looked and felt like, and as a practicing Quaker, he accepted an appointment as dean of studies at Pendle Hill, a Quaker community. There, he spent ten years honing his understanding and skills through living in a community built upon a faith in furthering one's authentic calling and "inner light."

Eventually, he felt that he was too far removed from the larger world and came close to taking a position as president of a college. Just in time, he convened a Clearness Committee, which is a Quaker process where a group of people who care about you asks open-ended questions to further your exploration of your life and decisions, without giving any advice. Palmer described to his Clearness Committee his wonderful plans with great enthusiasm, but when asked what he actually would like to do day-to-day on the job, he began reciting a litany of what he would not like: wearing a suit, raising money, showing up every day with time not being his own, and so on. Finally, it became clear to him that while the job was perfect in the abstract, it did not match how he truly wanted to live. However, in forgoing this opportunity, he also had to let go of the long-cherished dream of leading a whole organization.

While he was figuring out what to do, he told a wise friend that he was wondering when "way" (a Quaker term for the right path) would open for him, and he asked when it did for her. She explained that it never had, but "way sometimes closed," and she smiled as she reassured him that this worked just as well in clarifying what she was there to do—and what not. For Palmer, "way" did open, with a transformational impact on the world. As Palmer's life script was refined by his life experience, he began to focus on values-based education, and then he extended his reach to touch leaders more widely. Eventually, his work expanded so that he was guiding a movement—the Courage to Teach and the Courage to Lead—the ripple effects of which had a much greater impact than he might have achieved by following his original aspiration to be president of a college.

Another colleague and personal heroine of mine is an amazing leader who gained increased clarity through being unfairly made a scapegoat, which also produced a situation where she needed to retreat and regroup to refine her options for next steps. Martha Johnson was the Obama administration's first appointee to head the General Services Administration (GSA) for the United States government, a position that seemed like the midlife crowning jewel of a distinguished career. Within a year of assuming office, she had gained consensus to have this huge agency, which among other things is in charge of all government buildings and procurement, adopt a zero-environmental-footprint policy. In addition, she put processes in place so that decisions could be made more quickly and to ensure that new government buildings were more human-friendly and flexible than the faceless structures with long, identical corridors of the past. She also was popular with the GSA staff and with the administration.

However, when a scandal erupted related to exorbitant spending on a small conference about which she had no knowledge, Johnson recognized, in consultation with the administration, that she had to take one for the team and resign, even though the administration and her political party in Congress generally understood that she was not guilty of anything. As often happens (both sides do this), the other party was using this event to discredit the administration. She had enough Zeus savvy to understand that there was nothing personal in all this; it was just the way politics now works. And she had enough Demeter energy that through it all, her concern was for GSA, which allowed her to testify before Congress in a dignified and effective manner, with the goal of being sure the legislators and the American people knew what a good agency it was. The result was that she gained increased respect from her peers in government and from those in the know in the media.

Even with all this, what she went through was excruciating. For one thing, the experience of being made a scapegoat brings with it very archetypal fears of being exiled from the tribe, and for another, it generally releases a massive dose of cortisol, which can produce blinding fear, even when there is no actual threat to life and limb. Beyond that, the real bottom line was that she was out of a job she loved, and a major federal agency lost a great leader.

As I watched this drama unfold, I realized that there likely was another factor involved that I have seen with so many innovative and effective leaders. Their peers are threatened by their abilities, which are so stellar that they make others look bad, even though that is not their intent. This renders the higher-achieving person vulnerable, since others who would not have gone after her nevertheless will not risk anything to come to her defense. Getting results quickly in a government known for slow, deliberate processes can be viewed as disruptive of "how we do it here," just as it can be in other hierarchical, bureaucratic organizations. So, it is easy, when heads need to roll, to pick the head that is new, high achieving enough to be threatening to her colleagues, and loyal enough to the cause not to sue, and thus can be lopped off without a huge fuss.

Subsequently, Johnson had the courage to write about her experience candidly. Her book, *On My Watch: Leadership, Innovation, and Personal Resilience,* was designed not only to describe her ordeal, but also to help others integrate such painful experiences and move on. Johnson is now a consultant, helping government agencies, companies, and nonprofits operate in a twenty-first-century way.

Many leaders who are enough ahead of their time to create a backlash turn to consulting and coaching after being in senior management positions in places resistant to change. In such new roles, they can help a variety of leaders and leadership teams gain cutting-edge skills. As Jungian analyst and consultant Arthur Colman argues (in *Up from Scapegoating*), a consultant can help an organization see what needs to change, a reality that in turn creates backlash where many people want someone to blame for the discomfort they feel at having to do things differently. The consultant then can consciously play the role of ritual scapegoat when leaving the organization, freeing the leaders to make the required changes quietly over time, while the official blame is shifted to someone no longer with them.

What Parker and Johnson have in common is that both know who they are and do not substitute a job title for their identities, and each of them has a set of values deeper than shifting roles and circumstances. While both are people with a strong spiritual grounding, their values and principles are their own. We all can benefit from taking the time to clarify ours. I once came across a battered copy of a charming book by Joan Brady called *God on a Harley*. In it, God determines that the Ten Commandments in the Bible are just too generic to work for many people, as demonstrated by their inability to abide by them. So he appears to individuals disguised as someone they would welcome and helps them tailor the commandments specifically for themselves. In one instance, he shows up as a biker to get the attention of a young woman looking for a boyfriend. The exercise below is inspired by this book and designed to help you discover what principles are so core for you that they can help you weather whatever challenges life brings and hone your sense of self and purpose, even when this process feels painful.

Having explored how escalating challenges can require more complex abilities and a more refined sense of purpose, it now is time to move on to exploring how and why Zeus learns to be a more collaborative leader and the way in which this enhances his power while also enabling him to rule with greater ease.

APPLICATION EXERCISE:
Core Principles That Guide My Life

You might start by imagining someone showing up to help you with this task and what he or she would have to look like to get your attention. Then have a fantasy conversation with this deity, archetype, or person about where you might start in creating your custom-made rules for living. Have some fun making your own numbered list, perhaps including a few rules that you think should apply to everyone as well as some pertaining only to you and your strengths and purpose. To encourage doing this in a playful spirit, throw in a few lighthearted items, as well as deeply serious ones. If your religion has commandments, you might choose to include some or all of these, translated into language that is personalized for you and your situation. Once you have your list, think back over the times in your life when experience was chipping away part of you so that you could see yourself more clearly. Which core principles sustained you through these periods? Which principles would support you in any decisions you currently are facing or if such an event happened again?

Zeus Lesson Five:

Moving from Power Over
to Power With

THIS FINAL LESSON FROM ZEUS is offered to increase your ease and enjoyment in being the monarch of your life in a collaborative way. Having learned to take personal responsibility for fulfilling your various roles, with the help of this section you can move from the story that says, "It is all up to me," to a new story of "I just need to do my part and trust others to do theirs."

From his Greek culture, Zeus would have been exposed to many laudatory stories of heroes proving their abilities in battle and of rulers who were responsible for getting a kingdom on course by setting down decrees. Yet none of these stories adequately prepared him for a reign where he needed to satisfy diverse constituencies, to recognize how people feel when they are defeated and their old ways are undermined, to deal with the unrelenting stress of running things, or for Demeter's refusal to abide by his decision.

His early expectations for how social organization should work came from observing bees. Zeus's cave in Crete, far from his family, was a haven for them. He observed that they had a social system in which different kinds of bees performed different functions and communicated with one another to get things done. However, bees do not challenge their queen and do not get their feelings ruffled. They just do their jobs, as Zeus hoped people—and the other gods—would do. So while the bees gave him insight into how social systems could work, he had much to learn about governing other gods and people.

Zeus's strength initially was not in emotional intelligence, and his weakness in this area bit him in the back, first with the Titans and then with Demeter (and it does not appear that he ever worked things out with Hera). Moreover, managing the complexity of running things, with an opposition group (the Titans) always ready to pounce, was exhausting. As happens in times when an old story is dying and a new one is emerging, too much was going on for him to track it all.

At the outset, like many traditional people in power, Zeus focused almost exclusively on the gods he needed to please to keep their support. When Hades asked to marry Kore, Zeus thought it would solidify the bonds between the Upperworld and the Underworld and be a good political move. He did not understand Demeter's importance; his attention was more on political than on natural systems, and on masculine more than on feminine domains. Therefore, he did not consult her or her daughter. We can see the impact of a similar narrowing of attention in our time as well. In recent decades, our collective national attention has concentrated mainly on the horse race, the term used to describe the way the news media report virtually everything in international relations, politics, and sports, not only in terms of who won, but also, in almost every situation, who is gaining a bit of an advantage, like one horse getting its nose ahead of another's. This replicates Zeus's narrow focus on power dynamics. By crowding out complex analyses of what we can do about our own set of "famines"— that is, climate change, financial instability, caregiver fatigue, and so on—the emphasis on the horse race allows these problems to worsen as our attention is directed elsewhere.

In our personal lives, we might experience the result of such constrained attention when we have been so caught up in our work, and in completing everything on our to-do lists, that our life partners are feeling unloved or our children abandoned. Or at work, we may be so preoccupied with implementing a strategic plan in the context of heated organizational politics that we fail to notice a breakdown in the systems (like payroll processes, friendly and efficient customer service, or information technology) that support the whole.

When Hades asks Zeus for Kore's hand in marriage, Zeus, who is driven, distracted, and tired, just says, "Fine," agreeing with a hand-shake, never stopping to consider how Demeter or Kore may feel about this. It could be that Zeus is just too exhausted and busy to think the situation through fully. This proves to be very shortsighted, since it results in a famine that could end Zeus's reign and bring him to his knees once again. But as before, he learns, and so can the Zeus part of us.

At this major turning point in the story, Zeus does more than just course correct. He cannot succeed in solving this problem by follow-ing the old *power over* paradigm. He has to shift to *power with*. Zeus's breakthrough comes when he finally recognizes that his power over others is circumscribed. He cannot make anyone love him, but Aph-rodite could. He cannot make art, but Apollo can. He cannot make the crops grow, but Demeter can. And so on.

A successful resolution and healing come to Zeus when he discov-ers that capitulating is not the end of the world. Once he stops assum-ing that he knows best, he discovers how much wisdom and talent is all around him. Indeed, he experiences a true win-win as he rec-ognizes that he is not above the other gods, goddesses, and humans. Rather, they are interdependent, and Zeus's mandate is to coordinate and manage the others' efforts. The other gods did not turn on him, did not lose faith in him; instead, they just urged him to give in for their good and his own. Once Demeter allows vegetation to grow, humans again provide their offerings to the Olympian gods and Zeus's reign is secure once more.

A great weight is lifted from his shoulders when he recognizes the other gods' gifts and no longer has to feel responsible for everything. Because Demeter makes certain there is a good harvest, and all the other gods take care of their own areas of responsibility, he has less to worry about. This means he can be much happier and less stressed. But to get to this place, Zeus had to transform his story three times: first, from being an unwanted son in exile to aspiring to replace his father; second, from achieving his ambitions in a warrior mode to accepting the responsibilities of becoming the ruler of all the gods; and last, from having an autocratic leadership style to becoming a more collaborative leader. In each case, the transformation of his thinking occurred as a result of a crisis of some sort.

At some point in your journey, you may experience a challenge so different in magnitude that you have to change your whole paradigm or mind-set to meet it. This may be a sudden realization of some sort, or a new situation so context changing (the death of your spouse or partner, the collapse of your company, a diagnosis of a life-threatening disease) that it requires a reconfiguration of your psyche. Some people experience this as a kind of conversion (or epiphany) that happens quickly but takes years to integrate into the psyche; others may find that this happens gradually, as the immensity of the change settles in. The women's movement in the 1960s and 1970s was such a change, and even now men and women still are struggling to move from living very role-defined, and thus simpler (if less full) lives to managing multiple roles—in a society that does not yet completely support this shift and where there is not yet consensus even about manners (for instance, who opens the door for whom and when).

Many high-level managers now find that their tried-and-true methods no longer work. Leaders in all fields are discovering that they cannot just compel their workers to do what they want them to do but rather must give them input into decisions so that they are motivated to implement them. We see this in the change from assembly lines to "lean manufacturing," where worker teams have autonomy for every-day decision making. While strong command-and-control leadership used to be the norm (and to some degree still is in crisis situations), successful leaders and followers today demonstrate emotional intelligence, an ability to listen, a capacity for continual learning, and a facility to promote honest and productive dialogue and teamwork—all qualities that bring into play strengths that generally are regarded as feminine. This development reflects an evolution not only in humans but also in the expression of the Zeus archetype itself.

Our country's strength comes from its diversity, so it makes little sense to fight about whether to utilize or marginalize that asset. Republicans and Democrats need each other, since they each represent views held by American citizens. If they listened to one another, they might well be able to move past divisive wedge issues to a third story that allows our society to balance Demeter and Zeus in the service of actually addressing real, pressing twenty-first-century challenges.

I know that the tension goes out of my shoulders when I acknowledge all the support I have personally in my family and from my friendship network and work colleagues; when I see all the people besides me in the world who understand the dire challenges before us and are working on pieces of the puzzle; and when I watch so many amazing younger people who are sharp, articulate, and aware of the need for major change. We live in complex networks where our enjoyment of life (as well as our survival) depends on the work of others—those at the top of our social hierarchies and those lower down. And while many appear to be oblivious to the support we get from our government, if we pay attention, we will recognize that it gives us roads, bridges, and national and state parks; the rule of law needed for secure contracts; schools, police, and judicial systems; and so many other supports to the lives we live. And where would we be without the fecundity of the earth, which not only provides us with the raw materials for everything we have and do but also is a model for both cooperation and competition, encouraging us to keep both in their right places?

We are in the midst of a paradigm shift, from hierarchical thinking to ecological thinking. The association of one aspect of the feminine (Demeter) with the earth, which originally was so empowering, hurt women when connected with a hierarchical culture story. Aristotle came up with the idea of a hierarchy of animals, which actually conforms to modern notions about when they evolved, but he saw that hierarchy as just the natural order of things. He and other Greek philosophers did the world a service by developing the capacities of logical, rational thinking, but in the process of doing so, saw it as superior to the heart, soul, spirit, and body, which over time led to a belief that what was higher in the body (seen as the head) was superior to other parts. It is this thinking that historically has been retained, while the balancing wisdom of the Eleusinian Mysteries, which values all our parts, went underground. Thus, we can perceive that Zeus's granting Hades's request to marry Kore was his idea of a rational decision (it was a high-status marriage), while he would see Demeter's response as just emotional and maybe a bit hysterical. Moreover, during times when only men were educated to think rationally and objectively, women as a group came to be regarded as emotional, rather than logical. Their views could, then, be dismissed out of hand as irrational.

Medieval theologians carried hierarchical thinking further in the context of monotheism, developing the concept of the Great Chain of Being, with God and heaven at the top and Lucifer and hell at the bottom (with God often depicted as Zeus-like). In this context, belief in orthodox dogma replaced reason as a value. At the same time, *up* became synonymous with *good* and *down* with *bad*. By the time of the Renaissance, a social order of status and power was defined in detail, a pattern that influenced Shakespeare's dramas, even though actors fell very far down the list—below beggars. Such hierarchies contained certain dualistic comparisons, with men and masculinity associated with spirit, mind, rationality, and power, which were regarded as superior, and women and femininity with the body, the earth, emotions, sexuality, and the flesh, which were considered to be inferior and sometimes even sinful. All this was believed to validate male superiority to and power over both women and nature.

This connection between sexism and alienation from our planet is a legacy of cultural narratives such as these, which hang on in the unconscious long after few consciously believe in them. This helps to explain why many of us, as women, bristle at being associated with the earth, seeing that idea as insulting; carefully translate the wisdom of our hearts into rational terms in order to be heard; and avoid threatening men in power by speaking too authoritatively and seeming unladylike. At the same time, women find annoying the unconscious assumption many men seem to have that a man's subjective experience is the objective truth (an example: she says, "I'm cold," and a man who strongly supports women's equality still says, "It isn't cold"). Cultural stories remain alive in the unconscious long after our conscious minds know that they don't match the actual facts before us. Getting as close as possible to finding objective truth is important. No data supports the assumption that women are less smart or competent than men, yet we are paid less, and the glass ceiling only rarely is cracked. The best science we have universally stresses the importance of ecological sustainability, recognizing that humans have been living a story about our being the fittest that now needs to be questioned. This story is inconsistent with the reality that we are throwing the ecosystem out of balance and potentially threatening our own survival as well as that of other species. It is this kind of objective data, along with narrative scrutiny, that can help the Zeus archetype within and around us update our views.

A current, more nuanced and sophisticated study of evolution than the reductive, competitive "survival of the fittest" narrative is analogous to the shift in Zeus's management style. It requires recognizing multiple causality, with every part contributing to, and influencing, the whole. The cognitive complexity now required to analyze situations adequately is dependent on our questioning stories with contexts to see where they fit and where they don't. It also requires the generation of new stories predicated on transformed paradigms.

Because we still are living in the in-between of an incomplete revolution, concern about our planet is reflected in popular entertainment in ways that both reinforce the old reductive stories and showcase new, more complex ones. I frequently see trailers for what appear to be stereotypical apocalyptic science-fiction movies where some enemy arrives from outer space or under the sea and humans heroically save the planet by defeating this foe—with the world left in shambles in the process. While most of us love hero/villain stories, in this context, such plots can reinforce the erroneous idea that our problems are caused by some external enemy that needs to be killed or stopped. It is dangerous for us to be in denial about the human role in climate change. However, several very successful films express concern about climate change in a way that also portrays male Zeus-like characters who learn to honor Demeter's wisdom and do so in partnership with strong women. One example is *Avatar*.

At the time of this writing, *Avatar*, released in 2009, is the highest-grossing film of all time, suggesting that it hit a cultural nerve (plus it has great special effects). In an interesting parallel with Zeus's helplessness without his tendons (and perhaps his mental state when faced with Demeter's sit-down strike), the hero of the movie, Jake Sully, is a U.S. Marine who has demonstrated great valor but whose legs are now paralyzed from wounds he sustained in battle.[1] Scientists have bred avatars that combine genetic materials from humans and the Na'vi—the humanoid inhabitants of Pandora, a moon in the Alpha Centauri star system—allowing Jake, while lying in a special container with virtual communication capacities, to control his avatar's speech and movements.

Earth's resources have been ravaged, and mercenaries with military training are on a mining expedition to harvest a precious mineral found only on Pandora.[2] The leaders of the expedition are fully prepared to kill the Na'vi if they get in the way. Jake's job is to be accepted by the Na'vi and convince them to move to another location, out of the path of the scheduled mining. If he succeeds, this could save their lives.

In his avatar body, Jake meets Neytiri, a Na'vi woman, who saves his life. With courage and skill, she kills vicious beasts that are attacking him from all sides. He is impressed, and he thanks and compliments her, but she lays into him, telling him that he is stupid to make so much noise through his movements that he startles the wildlife. She does not see what she did as heroic; rather, she is sad that his naïveté caused all this needless death. She then teaches him Na'vi ways, which incorporate Zeus power with Demeter interdependence and care. The Na'vi represent Demeter's connected consciousness—with their planet and one another—combined with heroic courage and great strength. To Jake, the several scientists on the expedition, and clearly the viewer, the Na'vi seem superior to what humans have become.

With all of Pandora's natural creatures as allies, Jake's avatar helps the Na'vi drive the mercenaries away by sharing his knowledge of the invaders' plans and armaments and leading them into battle. Together, they save the Na'vi's way of life. At the end, the Na'vi connect their psyches with one another to produce a miracle, transferring Jake's essence from his dying human body into his avatar so that he becomes one of them and no longer is paralyzed.

This mythic movie's strange details serve as stand-ins (perhaps metaphors) for our ways of being. Jake moves out of paralysis when he lets go of an old stoic, macho ideal and seeks to live in harmony with others and with nature. Translating this into the terms of the Eleusinian Mysteries, he is not rejecting Zeus for Demeter; rather, he is integrating Zeus's courage and strength with Demeter's interdependent and caring way of being—a capacity that the Na'vi also possess. In so doing, he willingly risks letting go of the world he knew to enter the unknown. The implicit message is that you and I can move into an integrated Demeter-Zeus consciousness by first imagining the form it could take on our planet, in our individual lives, and then stepping into this vision.

The 2014 Intergovernmental Panel on Climate Change (IPCC) Report, prepared in partnership with the United Nations, was chilling enough in its well-documented dystopian vision of what is likely to happen, and soon, if we do not show more care for the earth. However, it is not so commonly recognized why male-female collaboration and feminine values are critical to achieving this. The preceding Demeter chapters provided some information on the efficacy of women's leadership styles. In *How Women Lead: The 8 Essential Strategies Successful Women Know*, Sharon Hadary and Laura Henderson summarize a number of studies that show that having a critical mass of women to balance masculine ways of viewing things produces more positive results than when few or no women occupy leadership roles. The findings consistently refer to having a high percentage of women leaders, not necessarily a majority of women or all women. They also do not demonstrate that having all women is the answer, any more than having all men is. The following are high points summarized on Hadary and Henderson's website and discussed in detail in their fine book. (The references in parentheses are the sources for their data.)

+ "Businesses with more women in leadership reported up to 69% higher financial results than those with fewer women leaders. (Pepperdine University)"

+ "Companies with the highest representation of women in top management achieved a 35% higher return on investment (ROI) and 34% higher total return to shareholders than those with the lowest representation. (Catalyst)"

+ "The most effective and collaborative teams have a greater proportion of women members. (MIT)"[3]

Although there is no way to demonstrate that the success of gender partnership in teams actually does bring both the Demeter and Zeus archetypes into the mix, I believe it is reasonable to assume that it does. The success of such teams can result partly from the different experiences and perspectives each gender brings to the table and partly from how a critical mass of women at the table encourages men to talk from both their Demeter and Zeus sides and makes women feel safer expressing both parts of themselves. This likely increases the complexity of the thinking in the group.

Success on the individual level also is enhanced by having internal Demeter-Zeus balance, giving us access to a fuller range of human abilities, rather than repressing those associated with the other gender, which used to be what was expected for healthy gender-identity development. The early research of pioneering scholar Sandra Bem revealed that men and women who are more androgynous (as measured through the Bem Sex Role Inventory, or BSRI, instrument) have a greater likelihood of achieving optimal adjustment to adult life, including finding success and fulfillment, than those who are defined rigidly by their gender roles. In today's work world, men are expected to exhibit greater tact with and care for employees, qualities that in general are traditionally regarded as feminine, and women are expected to show their tough, competitive sides, attributes still seen by many as masculine. Both men and women need to balance work with family responsibilities, which further encourages both sexes to demonstrate a range of nurturing and rule-enforcing capacities, especially when children are involved.

Bem's later research showed that people whose self-identity as a *person* is primary, with their gender identity being secondary, pick what they think, like, and do from a wider range of human attributes and preferences than those for whom gender identity is primary. They are less likely to stereotype others based on current ideas about gender in the culture.[4] Today, gender and sexuality often are seen as being on a continuum that includes not only culturally defined qualities associated with men or women but also biological ones such as sexual orientation. Yet the feminine and the masculine still carry enough historical weight that they remain meaningful concepts.

The androgynous advantage has become so apparent for both men and women that it is now a part of some leadership training designs. Leadership scholar Jean Lipman-Blumen interviewed five thousand leaders and integrated what she identified as male and female management styles in a holistic approach appropriate for our time. In the resulting book, *Connective Leadership: Managing in a Changing World,* she argues that global interdependence calls for greater collaborative abilities than were needed before, and she notes that these qualities are associated with a feminine leadership style. She also asserts that, while essential, an emphasis on collaboration is not a substitute for the continuing need to recognize that companies, countries, and other entities still see themselves as independent and are looking out for their own competitive interests, motivations that remain associated with a more masculine leadership style. As with moral development theory, Lipman-Blumen found that these leadership styles do not reflect absolute male-female differences, since there is substantial crossover. From her interviews, she identified a wide range of collaborative and competitive leadership attributes and then created an instrument to measure them. Her work is now widely used in assessing leadership competencies and helping men and women enhance their personal strengths, with the full range of human capacities relevant to fulfilling their responsibilities.

Having explored the transformation within the Zeus archetype from its more primal manifestations to its more advanced and contemporary forms, generally integrated with the Demeter archetype, it is time to move on to consider Persephone and the way of transformation.

APPLICATION EXERCISE:
My Zeus-Demeter Integration

Create a drawing or a chart depicting yourself as if you had a Zeus side to your body and a Demeter side. On this diagram, draw symbols, paste pictures, or write words that describe your Demeter and Zeus qualities. Put the ones that you see as your qualities inside the figure and those you rely on in others on the outside, placing them as close to you or as far away as seems accurate. Then add lines between you and these external helpers that reflect the quality of your relationship. For example, a straight line might suggest a relationship that is direct, focused, and action oriented, while a wavy one might indicate one that flows, or a dotted line for one more tentative. (If this seems too difficult or time-consuming, just imagine such a drawing.)

CAPSTONE EXERCISE:
Dialogue with Zeus

In a journal or notebook, dialogue with Zeus where you speak as yourself and the Zeus of your imagination speaks for him. Start by thanking Zeus for how he has enhanced your life, then ask how he would like to be expressed in you. You can then say what you would like from him, including where you think he would be helpful and where not. Allow a natural dialogue to develop wherein you come to an agreement about Zeus's future role in your life. End on a note of appreciation.

Part Three

PERSEPHONE

Persephone and the
Way of Transformation

THE RAPID PACE OF TECHNOLOGICAL CHANGE in our world requires an unprecedented capacity to adapt to shifting conditions. New fields pop up and old ones disappear at an amazing rate, and what is required to succeed in our work and personal lives keeps escalating. Global interdependence creates a situation where economic crises, natural disasters, environmental accidents, and political upheavals in one part of the world affect those in another, sometimes changing the context in which we live (e.g., fewer jobs, more expensive gas, or a crisis taking active duty military and National Guard members to dangerous places).

How do we respond to this continually changing landscape without being chameleons and losing touch with who we are in the process? How can we have a secure sense of our future when it is difficult to predict what will happen tomorrow, never mind a year from now? How do we change our views in the light of new information and situations? How do we hold fast to values that seem essential to us in a context where others do not necessarily share them? And how do we avoid being so taken over by our multiple roles and tasks that we lose ourselves?

Just as Zeus helps us to develop healthy and strong egos and Demeter helps us to open our hearts, Persephone helps us to connect with the deeper part of ourselves, something essential within us beneath our roles or learned preferences. It is here that we find an identity solid enough to allow us to change to meet new conditions, and to utilize the full range of our gifts. We discover this deeper part of us by determining whom and what we love—not so much whom or what we want to care for, but what lights us up. The ancient Greeks were more attentive to the nuances of love than we commonly are today, leading them to develop different words for different types of love. For example, the word *philia* meant "friendship" or "sibling" or other love between peers. *Storge* referred to the natural love of parents and children for one another, and sometimes between rulers and subjects. *Agape* is a spirit of charity toward others, and in Christian usage, God's love for humankind that we are challenged to emulate. All of these kinds of love are embodied in the Demeter archetype. Persephone, however, personifies the wisdom as well as the challenges of *eros*, which is a Greek concept often misunderstood or only partially understood today.

The word *eros* refers to sexual and other forms of intimate loving and, according to Plato, provides us with an appreciation of spiritual beauty and truth that can lead to transcendence. Eros also is associated with the life force and helps us to choose aliveness in our work, as well as in our personal relationships. Psychoanalyst Sigmund Freud associated eros with the drive for love, creativity, sexuality, and personal fulfillment.

Sadly, the word *eros* is now most widely known through the word *erotic*, which generally is used to describe "exotic" dancers or even pornography, neither of which are true expressions of it. We can be turned on by a male or female stripping or by encounters with anyone who flaunts their sexuality or flirts with us. That is an impersonal, sexual response. Eros, however, is highly personal, even individualistic. The personal nature of eros explains how, although you might be attracted to numerous people, you know you want to become romantically involved with only one of them in particular, and perhaps even make a lifelong commitment. Although it sometimes envelops us with longing, the call of eros can be as gentle as a spring breeze or as fragile as new sprouting grass. It is easy to miss it or trample on its first manifestations, thereby killing it. When we do this, our lives can appear to be fine and our choices sensible, yet feel strangely unfulfilling.

You know you are experiencing eros when it calls to you from some deep inner knowing, something your ego self would not have chosen. In this way, it serves as a sort of GPS from the soul, helping us to intuit where we need to go next by what attracts us.[1] Freud contrasted Eros (the Greek god of love who shoots the arrows that determine whom and what we love) with Thanatos (the god of death, which he saw as connected with aggression, violence, and sadism). However, in Persephone's story, *thanatos,* as Freud defined it, emerges as complementary to eros, not its opposite. In its most positive form, thanatos (in psychoanalytical terms, the death instinct) provides us with the ability to die to what we have been so that we might be resurrected to what we need to become, just as Kore dies to being a child and, as Persephone, becomes a wife and a queen. That is why Persephone achieves completion in marrying Hades, the god not so much of death itself, but of caring for the dead. Together, through their union, Persephone and Hades support transformational processes.

However, thanatos's shadow side remains as Freud saw it, as the drive to exterminate ourselves, which appears quite active in the epidemics of addiction and suicide and the appeal of martyrdom to suicide bombers, as well as its collective expression in nuclear proliferation, environmental devastation, and genocide. In Jungian thought, and in many spiritual traditions, the cure for the toxic potential within an archetype is its positive aspect, which in this case is the ability to let go and move on. The overall importance for lessons from Persephone today comes from this need to be true to yourself and stay connected to others and the world while also remaining relevant to a world that is changing at such an accelerated rate that it requires us to die regularly to what we were so that we can be transformed into what we now must become.

The Archetype of Persephone

Persephone is portrayed in ancient Greek art as a very beautiful, dark-haired young woman—so beautiful that there are stories in Greek mythology about virtually all the male gods wanting her. Her nature is so mysterious and complex that she is simultaneously the queen of the Underworld and also the goddess of spring. Persephone's ease in moving back and forth between the worlds and the seasons can be a model for our gaining ease in shifting between multiple roles and adjusting to new life stages that require different things from us.

In early and later versions of the Eleusinian story, Persephone commonly is depicted visually and in written descriptions as (1) emerging from the Underworld, (2) stepping into the Upperworld with flowers blooming as her feet touch the ground, or (3) standing with great dignity with her right hand outstretched to warmly welcome initiates into their lives after death. She also is shown in mutually loving scenes with her mother and separately with Hades, the lord of the Underworld, suggesting that she integrates, or at least is comfortable with, their very different mind-sets.

Where Demeter taught initiates how agriculture, life, death, and procreation work, Persephone's story dramatized it, so that initiates could identify with her experiences as ones they also do and will live. As a goddess of transformation, Persephone is seen metaphorically as the seed planted in the ground. Seeds germinate in a way that makes them appear to be decomposing before they sprout. Similarly, the caterpillar that goes into the cocoon melts into liquid before it reforms as a butterfly. The plant in fruition, the butterfly, and the optimal you are going through a process that is fully authentic and natural, even though, in each stage, it looks different and can feel pretty scary.

These classical visual images of Persephone are laden with symbolic meaning. Persephone sometimes is shown holding barley, as at the closing of the Mystery rites, and a snake. In the context of renewal, flowers, barley, and snakes all fit together. Barley evokes Demeter wisdom of the planting and harvesting cycle, which Persephone continues to uphold; both flowers and snakes suggest renewal (flowers signifying spring, snakes annually shedding their skins); and flowers conjure up vagina images (as in many paintings by Georgia O'Keeffe) and snakes, phallic ones. Persephone's wisdom helps us become intimate with ourselves and then with others; thus, she is regarded frequently as the Underworld equivalent of Aphrodite, the goddess of sexuality and romantic love. People often learn the way eros works first in their romantic life and only later in other aspects of it, including professional and spiritual callings.

The Persephone archetype can seem paradoxical from a modern perspective. We live in a culture that equates dark with evil, and light with good, and that typically thinks that deep spiritual wisdom should be talked about in quiet, pious tones, even though we see pictures of the Dalai Lama always laughing. In her Underworld role, Persephone generally is viewed as a goddess who embodies the dark feminine, which is both wise and lighthearted. It is important here to rid our minds of any idea that darkness, in this context, is associated with evil.

Persephone is a dark goddess because she presides over realms where things happen that are private or secret—in the afterlife, in quietly shared confidences, in desire, and in sexual union. In the Middle Ages, the orgasm was even called the "little death" because its bliss takes one out of the ego part of the self that wants to control life rather than experience it, and also because eros and death remain irrevocably linked. Persephone also stands for all that is invisible about ourselves to an external observer: what one is thinking, consciously and unconsciously, the experiences of dreams and imaginative fantasies, and moments of creative inspiration or vision when someone is touched by the muse. Most importantly, Persephone represents the consciousness of the deeper self that most people are in the dark about and that speaks to us in nighttime dreams and visions. This explains why the secret parts of the Eleusinian rites, held indoors in the Telesterion, revealed truths beyond those shared in the well-known story of Demeter and Persephone or the parts of the rites that were performed in public view.

Marion Woodman and Elinor Dickson, in *Dancing in the Flames: The Dark Goddess in the Transformation of Consciousness*, explain that the words *dark* and *wise* are almost indistinguishable in ancient Semitic script. Examples of other dark goddesses include the Black Madonna, the statues of which speak so powerfully of mysteries beyond words, and the Gnostic Mary Magdalene, who was described in the Dead Sea Scrolls as the leader of Jesus's disciples and credited with having greater understanding of his teachings than the others. Some even think she was Jesus's wife.[2] According to biblical scholars Susan Cole, Marian Ronan, and Hal Taussig, in *Wisdom's Feast: Sophia in Study and Celebration,* the ancient Hebrew goddess Sophia, whose name means "wisdom," was God's partner, was with him at the creation of the universe, and delighted him with her astuteness and lightheartedness.

Persephone similarly maintains an attitude of lightheartedness, which may be why the Eleusinian procession from Athens to Eleusis included dancing and jesting, some of it ribald. Moreover, dark goddesses are known for "childlike energy—spontaneity, play, creative ideas,"[3] and approach life with grace and ease. We can see this in Persephone's maintaining some of her childlike Kore nature as well as having the shamanistic ability to move between the worlds, at least after she has gained free access to both the Upperworld and the Underworld. In modern times, such an ability may take a variety of forms. It can be as simple as moving comfortably from situation to situation, even those that are foreign to you. It can mean showing resilience when faced with life's setbacks and difficulties, as well as flowing with the challenges of adapting to the ways of the people you love. In addition, it includes going back and forth between the outer and inner lives and between the conscious and unconscious minds.

In this way, Persephone connects us with our deeper selves, or souls. From the psychological perspective, the soul is a deeper part of us than the ego, mind, or even heart, and a more reliable source of guidance about what is right for you or me than the conventional morality of any particular time or place. This psychological meaning does not imply immortality or a relationship with a supreme being, although it does not preclude either. In Christian usage, the soul is the eternal part of us.

If we are fortunate, our souls, hearts, egos, and what others want from us will be in alignment when we make major choices. But for many of us, much of the time, this is not the case—hence the many literary, artistic, and musical works about the archetypal connection of love and death, eros and thanatos, and what happens when social roles and expectations conflict with true love. In the context of their warring families' unwillingness to allow them to marry, Romeo and Juliet die, choosing eros over their social and familial roles. And in our own lives, we see the converse, with so many people who have died inside because they rejected the calls of eros.

The association of eros and death is essential to Persephone wisdom in other ways as well. She was Apollo's partner in fostering the practice of incubation in temples throughout Greece and that was included in the Eleusinian pilgrimage. Persephone urges us to learn to "die before we die" so that we are free to trust life. Incubation was seen as connecting Apollo above and Persephone below, perhaps suggesting that he could help shine light on her dark Mysteries. People would be swaddled and placed lying down in private spaces, where they moved into a trance state, such as when you are falling asleep or waking up or when you are half asleep in the middle of the night. It is a state of consciousness where answers may come to you or dilemmas you have been avoiding press for attention. The hope was for participants to have a vision that would help them experience the journey to the Underworld and what it might mean for them. Scholars such as Peter Kingsley (*In the Dark Places of Wisdom*) tell us that in incubation, the ancients had visions of what the afterlife was like; thus, they experienced this transition vicariously so that they would not fear it when they had to say good-bye to their bodies.

Imaginatively dying before we die today also can prepare us to accept the little deaths we experience every time we let go of some idea, attitude, capability, person, place, job, illusion, dream, and so on. Stephen Levine and Ondrea Levine, in their seminal book *Who Dies? An Investigation of Conscious Living and Conscious Dying*, describe how responding to all these little deaths with grace and awareness prepares us to meet our actual deaths with equanimity, as Persephone seems so able to do.

Although Greek myths sometimes referred to the Underworld as gloomy, the versions that stressed Persephone's cooperation with Apollo always spoke of light in the Underworld, just as there is dark, at night or in shadows, in the Upperworld. Apollo, as the sun god, was believed to make the rounds through the heavens during the day and to complete the circle in the Underworld at night, so the sun that shone on high at midday also shone in the Underworld at midnight.[4]

The Greeks were not dualistic; they considered light and dark more like yin and yang, complements rather than opposites. The most secret parts of the Eleusinian rites were held in complete darkness, until a moment when a sudden great light appeared so that initiates felt as if they were experiencing midnight in the Underworld, and by experiencing the wisdom within the darkness and the light, they felt ready to be reborn. In that way, they were seeing that there was life in death and death in life, all related to the cycles of birth, death, and rebirth.

The Underworld in our psyches often is regarded as a symbol of the unconscious, part of which is the *shadow* and part of which is our deeper selves. As you know, this contains the negative potential of each archetype that it is best we not live out. Even Persephone's flexibility has a shadow side. When we call someone a "real snake in the grass," we are referring to the archetype's negative potential. The fact that Hermes shepherds Persephone back from the Underworld may be a reminder that her shadow side is coming with her, just as ours always goes with us. (This is similar to how Zeus being married to Hera reflects his unhappy feminine side, or Persephone being married to Hades reveals her deeply passionate nature and her relationship with death in both its negative and positive forms.) Because he is so mercurial (as a Roman god, his name is Mercury), Hermes has the ability to morph into what he needs to become to meet new circumstances or get what he wants. The negative manifestation of his archetype is the person who serves up charm laced with lies for personal gain, which would reflect the potential shadow side of Persephone's attractiveness and flexibility. He also can be expressed as an Iago kind of character (from Shakespeare's *Othello*), who uses trickery simply to harm another, for revenge or for no apparent reason, as well as by the prostitute or chameleon, who will be whatever you want for enough money, power, or even safety, and who frequently may be victimized. When Persephone emerges with Hermes, one of her first acts is to lie to her mother to avoid her wrath (a typical adolescent thing to do), saying either that Hades tricked her or that he forced her into eating the pomegranate seeds.

This is what can happen to the Persephone part of us when we have her flexibility but never get centered in a deeper soul identity. Some Persephone types fall into the trap of being people pleasers, manipulators, even con artists. If, in addition, they lack the strength of Zeus and solidity of Demeter, they may not have the boundaries needed to stand up for themselves, so they manipulate others to do that for them. Holding on to a victim identity, they always seem to need rescue.

The shadow within, however, is not only where we hide our negative attributes; it also is the place we bury the positive potential that we are unaware of or that does not fit into the mind-set of our time and place. For much of the recent past, many of the facets of the Persephone archetype have gone underground because of being denied, undervalued, or demonized. Ironically, what has remained is her association with sexuality and eroticism, whose negative shadow form is apparent in the amount of pornography on the Internet, and sometimes an association with images of a seductive but evil witch.

Given this, we can think of the positive qualities of the Persephone archetype as hidden treasures that, when integrated into our psyches, can transform our lives. Allowing repressed gifts and creative insights to see the light of day releases energy that can put a spring in your step, so you feel as though you are moving from winter's gloom into spring's glory.

Persephone's Stories

Because Persephone's wisdom and perspective were shared only in the secret parts of the Eleusinian rites, those writing about them (then and now) inevitably project their own perceptions of her story onto it, viewing her through their own lenses and in the context of the situation they are using the story to explain.

The Classic Story is always some version of the myth that began this book, with variations that present Persephone more as a passive victim than I do, or Homer does. In some, she not only is abducted, but also violently raped by Hades (a story sometimes told with rather off-putting salacious enjoyment).[5] She is miserable in the Underworld until rescued by the efforts of her mother, but because she was tricked into eating some pomegranate seeds, she is destined to suffer several months of every year in this dark and dismal place, stuck with the man who abused her (even Homer reports her saying she was tricked into eating the seeds). It seems to me unlikely that this Persephone could have been the heroine of the Eleusinian Mysteries, but this story has been utilized in very healing ways by depth psychologists.

The Depth Psychology Story: In it, the descent to the Underworld is used as a metaphor for the process of healing. Persephone's story of surviving and recovering from rape then can be of comfort to women who have experienced sexual and other abuse, in the same way the story of Demeter's experience of loss soothed me. Somehow, recognizing that you are not alone and that people throughout history have endured the same thing can take away a bit of the sting. When psychologists use the story in this manner with clients, they generally move on to interpret it as a journey to the inner world. Persephone's going to the Underworld thus becomes a metaphor for examining one's inner life, which includes facing the traumas and difficulties you have experienced, whatever they might have been, and becoming aware of your own shadow, so that you can heal and then find your deeper self. Many people who are knowledgeable about depth psychology will refer to having been in the Underworld as a shorthand for having focused inward, whether because of a trauma in order to heal, because of a turn in the road that requires them to stop and regroup, or as a conscious choice to go inward to attain greater self-awareness.

The Metaphysical Story: At the spiritual level, the overall archetypal pattern undergirding Persephone's story is that of birth, death, and rebirth (or resurrection). As Timothy Freke and Peter Gandy rather provocatively note through the title of their book, *The Jesus Mysteries: Was the "Original Jesus" a Pagan God?*, stories of a god who dies and is reborn and whose resurrection provides hope for life after death for humans were common in Mediterranean and Near Eastern mythologies during the formation of Christianity and the Eleusinian Mysteries. Though Freke and Gandy focus on male gods who embody this pattern—Osiris, Dionysus, Attis, and Mithras, all of whom died and were resurrected—we can add Persephone, since her being abducted to the Underworld and reemerging is another kind of death and resurrection. The teachings associated with each of these male god figures are too divergent to substantiate the idea that they all are versions of Jesus. For example, Jesus's teachings emphasized love for one's neighbor and the power of forgiveness, while Mithras reinforced a more warlike consciousness. However, all of these male figures are killed in cruel ways and are reborn, pieced together, or resurrected, an archetypal pattern also reflected in many indigenous shamanic initiations.

What differentiates the archetypal pattern of death and rebirth in Persephone's story from the others is that she did not experience a dismembering death. Her archetype provides a path for mortals to undergo spiritual transformation by following eros with an attitude of learning to understand and trust the processes of life and death. Because many ancient Greeks, including Pythagoras, Socrates, and Plato, believed in metempsychosis—reincarnation, or the transmigration of the soul—Persephone's journey could be seen as enacting the recurring cycle of life, death, and rebirth.

According to Plato (in *Republic,* especially The Myth of Er), the Underworld was the place where the newly dead, having left behind their physical form, were stripped of everything else—their egos, bodies, minds, and hearts—leaving only their souls behind. After they were thus purified, they engaged in choosing another form for their next life. There were various beliefs floating around in their time as in ours about what that really meant. You could be reborn into earthly life or go to some other realm that fit your level of spiritual attainment. Then and now, metempsychosis could be understood as being entirely about this life on this earth and the death and rebirth experiences we can undergo at any time of life. In all of these options, it was assumed that soul development and refinement would stand us in good stead for our next steps in this life, in the next one, or in some afterlife.

The Prepatriarchal Story: Charlene Spretnak, in her essay "The Myth of Demeter and Persephone" (in Christine Downing's *The Long Journey Home*), argues that in prepatriarchal versions of Persephone's story, it would have made little sense for her father to have given permission for her abduction, since such patriarchal customs evolved later. In this alternative account, Persephone simply volunteers to go to the Underworld because she witnessed the confusion suffered by its inhabitants, who had no way to understand what was happening to them, and feels a vocational calling to help them. This would explain the fact that she was known as the queen of the Underworld long before there was a myth that described how she got there. This version highlights Persephone's journey as a mission of mercy—an altruistic love story for those who need her.

A Coming of Age Heroine Narrative is the primary version of the story that provides a frame for part 3, but aspects of all these other stories also are present when they are helpful in awakening the heroine within. Part 3, then, explores the Persephone archetype as it helps us follow the path of eros in life choices related to love, work, identity, mortality, and metaphysics.

- ✦ Lesson 1 explores the role of eros in developing your ability to balance independence with intimacy.

- ✦ Lesson 2 continues this theme with a particular focus on the challenges of the teenage years and how it is never too late to address them.

- ✦ Lesson 3 provides direction for reconnecting with eros as an inner guidance system when feeling lost.

- ✦ Lesson 4 explores how the capacity for discernment and making wise choices can help you realize your destiny, even when you have been abducted by a fate you would not have chosen.

✦ Lesson 5 considers the metaphysical narratives that guide our lives and how we can be happier by trading in stories that breed despair and alienation for ones that offer hope and the feeling of being at home in the world.

The Persephone Archetype

Mythological Persephone: the queen of the Underworld and goddess of springtime

Primary Heroine Lesson: trusting life and your own ability to handle whatever happens

Narrative Progression: from naïve trust to skilled transformational abilities

Gifts: lightness, flexibility, creativity, inspired choosing, and ease with partnering

Historical Gender Association: the feminine, able to partner easily with the masculine

Decision-Making Mode: intuition combined with well-honed discernment

Inner Capacity Developed: connection with the deeper self (soul)

Counterproductive Forms: pleasers, prostitutes (not just sexual), manipulators

Persephone and the Eleusinian Promise

Primary Happiness Practices: optimism, letting go and moving on, lightness, and humor

Contribution to Prosperity: intuition, inspiration, creativity, and game-changing innovation

Contribution to Freedom: adaptability, acceptance of death, and trust in inner guidance

APPLICATION QUESTIONS:

Based on what you know so far about Persephone,

Do you have too much, too little, or not enough of this archetype?

If it is present in your life, was it a family legacy (i.e., you were taught to be like that and are), a vocational ally, or something deeply and authentically you?

Do you like people who reflect Persephone's stance in the world? Why or why not?

Persephone Lesson One:

Responding to the Call of Eros

THE FIRST AND MOST IMPORTANT LESSON the Persephone archetype offers you is the ability to know who you are at a much deeper level than your ego or even your heart. The benefit of such knowledge is that you can become more intimate with yourself and thus have the capacity to experience greater intimacy with others. To aid you in this process, this first lesson in part 3 will begin with Persephone's story in the context of ancient Greek practices and beliefs, then of the romance genre of fiction, and finally of psychological findings about love and human happiness.

In ancient Greece, girls soon to be brides traditionally went to a meadow with their friends to pick flowers as preparation for marriage (sort of like a bridal shower, except there were flowers instead of today's gifts). Most often, they married a life, not just a person, with a ready-made job as a homemaker and a ready-made home, since initially, a wife likely would live in a household with her groom, his parents, other relatives, servants, slaves, and so on. In the era when practice of the Mysteries was at its peak, a traditional wedding ceremony would include a mock abduction, in which the groom grabbed the arm of the bride. Folklorists see this tradition as the precursor to the later convention of the groom carrying the bride over the threshold of their marital home. All these signs suggest that the myth of Hades's "abduction" of Kore may have been symbolic only, informed by existing cultural practice, and that she may have been, by the standards of the time, ready for marriage.[1] In the modern world, an analogous archetypal Persephone narrative—the coming-of-age love story—is alive and well in romantic novels for teens and adults but updated to reflect the very different and freer realities of women's lives today. This lesson builds the capacity to be attentive to eros when it calls you, and also to treat it with discernment, since its messages often are subtle and nuanced.

When Kore picks the evocative narcissus, the earth moves, the ground opens up, and a handsome dark stranger rides in on a chariot, sweeps her off her feet, and carries her away. We can look at this two ways: Consciously, she is just picking a flower. Subconsciously, she has a desire she does not even know how to name. Most men and women love the feeling of being swept off their feet by mutual infatuation. When my now husband, David, and I first shared an office, neither of us acknowledged our attraction to the other, while the very air in the room gradually became charged with the unspoken. He sat on one side of the room with his feet up on his desk, looking very fortress-like, and I sat quite primly behind mine as we had heated discussions about literature, a subject about which we greatly disagreed. Then, late one afternoon, I remember finding myself standing next to him with no recollection of how I got there; it seemed as though I had just floated to his side. He greeted me with a look of combined terror and joy, as we both faced the reality that we had been abducted by love. There clearly was no turning back, but neither of us had this in our plans, so it felt like it was turning our lives upside down. And, very soon, as you already learned, we were living through tragedy together while also grappling with the multiple roles and challenges of modern life.

As with my experience with David, although Kore's was unsettling and unexpected, it does not suggest that Hades's appearance was entirely unwelcome or ultimately wrong for her. Bad boys appeal to women, especially young ones, and at any age, there is a sense that bad-boy sex might be more fun than good-boy sex, as long as the woman actually feels safe. In truth, boys we think of as bad are not necessarily evil, but they are energetic, courageous, and lusty.

While Hades is your classic "dark man," whose lure is partly a sense of his transformational qualities, he was not seen within Greek culture as a negative god. The dark masculine is similar to the dark feminine in being frightening because its appeal takes you into the unknown. Although Hades is the lord of the Underworld, he does not kill any-one. It's the Fates who determine the time of one's death and Thana-tos who executes this plan. Hades cares for the dead and oversees the process of their renewal, metamorphosis, and transformation, so that a being can go on to the place or the challenge that comes next.[2] How-ever, as you already know, the Greek idea of what that would entail does mean he presided over divesting people of their bodies, egos, minds, and hearts, leaving only their souls.

Hades, and even more so his Roman equivalent, Pluto, is associated with prosperity, related particularly to the underground pressure in the earth that creates precious metals, jewels, and oil, as well as the process of seeds germinating. By analogy, he is responsible for purify-ing the consciousness of the dead, so that they let go of the trappings of their last life's personality. As the dark man, he also is the lover who penetrates into one's soul essence and wants to know the deeper, real you, not just your sparkly persona. By extension, he can assist you at any of life's major transitions, when you need to move from being a child to an adult, from a young adult to someone in midlife, from your working life into retirement, and when you are approaching and then facing death.

Adolescence is a death and rebirth process, whether in ancient Greece or today, and it is a time of grappling with issues of sexuality and when many teens court death by doing dangerous things. The appearance of Hades is an announcement that it is time for Kore to shed her girlish identity and move on to the next stage in her life. How much self-esteem a girl has often can foretell how able she is to keep her head enough to minimize damage to herself or others in this time of major transition.

We do not know what happens when Kore first finds herself in the Underworld, but given her status and prestige (as Demeter's and Zeus's daughter), it is unlikely that she would take kindly to being forced into anything. However attractive Hades may have seemed in the meadow, she would hardly be prepared to all of a sudden be in a world she does not know. And even if she were told that they were to officially marry, and she were to agree before she actually feels ready for it, she would not have forgiven a rape. She is a girl used to playing in the meadow, surrounded by flowers, accustomed to beauty and a sense of being adored in community. She certainly would miss her mother. And she would start to worry: "Who *is* this man who seems so debonair? He presides over the dead, for goodness sake! Scary!"

Just as Zeus is associated with the stance of the ego, Hades is associated with the id—pure, unsocialized desire. Therefore, he would have been impatient to marry Kore and might have assumed that all he needed was Zeus's consent to just go get her. Crucially, however, Aphrodite had inflamed Hades with *love* for Kore, so his passion was not just lust. Thus, we could imagine that, after the abduction, his better sense kicked in and he courted her. New to the Underworld, she may have felt like so many women swept off their feet by a new husband and then suddenly confronting the dreariness of being a housewife in a home that really is not that nice. Even so, as a typical "good girl," it would make sense that Kore initially would follow her pattern of pleasing and doing what was expected. And she was her mother's openhearted daughter. She cared about the people there. Her caring moved her into her role as queen, helping those in the Underworld to have a better quality of experience because they knew where they were and what was going to happen. Even today, we begin to know our purpose by the desires of our heart—not just by what we want to do for others, but by what we want to experience ourselves.

The Persephone archetype informs much of romance fiction, written for women of all ages, whether or not the author even knows this myth. I suspect that this genre is so popular with women because it generally gives them permission to live a love story and reaffirms that love can triumph in the end—a message much needed in our society. If you visit Goodreads.com to look for books explicitly based on the Persephone story, you will see a list of novels that goes on and on and on. The Persephone archetype also is ever present, even when not identified by name. In the stereotypical plot, a romantic but dangerous-seeming dark and mysterious man sweeps a beautiful heroine off her feet (usually, but not always, metaphorically). Her predictable task is to discern whether or not to trust him and her attraction to him. Literature also provides us with many plots wherein the male hero encounters a dark-haired, seductive woman, experiences a similar sense of being overwhelmed by desire, and also needs to discern whether becoming involved with her is dangerous or not. Typical vampire plots go further with their reliance on the connection between eros and death that is so attractive to readers, especially adolescents.

What is typically considered women's fiction—romances, mystery stories, and fantasy—accounts for a huge percentage of book sales, and most iterations of the genre include a romantic plot or subplot. Yet these books are disdained in a way that hero stories, especially those written for men, are not, although both genres range from very bad novels with stereotypical plots and terrible prose to well-written and psychologically astute creative literature. The particular undervaluing of the romance form in women's fiction is interesting, not only given the importance of romantic love to individual happiness and family and societal health, but also because these stories provide the main training many of us receive in discerning how to follow eros in any or all of its forms. Men write about the complexities of love, too, but when they do they often are lauded for writing well about life.

What I notice about love stories—classic and contemporary—is that often the soul of the heroine speaks through visceral yearnings while her mind stays wisely cautious. That is why lovers in classic works of romantic fiction often miscommunicate, leading to misunderstandings and conflict. Contemporary romance novels of some psychologi-

cal and literary merit also have much in common with Jane Austen's
Pride and Prejudice, which could be the poster book for the intelligent
love story. Elizabeth and Darcy are attracted to one another, but she
believes he is an arrogant snob (and he reflects more of the Zeus than
the Hades archetype, even though the latter is more prevalent in
love stories). Darcy fears, as does his family, that Elizabeth is a social-
climbing gold digger. He overcomes his pride, and she her prejudice,
as they learn more about one another. When their judgment confirms
the wisdom of their attraction, they marry.

Such processes of caution and discernment are critical to thriving, not in romance alone but in all aspects of life, since they provide
imaginative practice, through identifying with the characters, for how
you can follow your desires without losing your head. This book could
track love story after love story, but you know the plot, so there is no
need. The fulfillment of underlying longing for union with another (at
the root of these love stories) is an important part of life for many of
us. So let's turn from literature to real life. The Persephone archetype
helps us to find our deeper selves and, as a result, be able to be intimate with others—at appropriate levels for different kinds of relationships. Accomplishing this is a—perhaps *the*—primary task of human
maturation.

Elements of how this works out for us are imprinted by our first
love—our experience with our mothers. We begin life in her body,
and when we are born, we are held in her arms. If we nurse, we continue to be fed from her body while being enfolded in an embrace.
Deep bonding occurs (ideally) as our eyes meet hers and we learn
the first elements of being human by mirroring her smiles and other
expressions. Psychologists call this entrainment, an intimate relationship of mutuality that prepares us, according to attachment theory, to
be able to bond with others throughout life. If we grow up optimally,
we gradually develop a separate sense of ourselves, supported by this
maternal bond. Young boys and girls, even today, will say they want
to marry their mothers, as Freud likely had observed in his time, but
when puberty hits, we now know that they become attracted to those
their own age and of the gender determined by their sexual orienta-

tion. Having had a strong, loving, and healthy relationship with their mother actually fosters the ability to bond with a lover and commit to a life mate. The physical intimacy of being lovers fosters a more mature form of entrainment. Committed lovers share a bed, know each other's bodies inside and out, and, ideally, benefit from the emotional intimacy of sharing their feelings, hopes, and dreams. This entrainment is reinforced continually and thus strengthened with life partners if they care for each other through events such as childbirth, in illness, or when emotionally devastated. In such adult relationships, it is important that lovers each keep aware of their personal identities separate from one another, just as it is important for children to differentiate from parents while also remaining dependent on them. Otherwise, lovers can get so caught up in being a couple that they lose their individuality. When that happens, the psyche generally pulls away and the erotic connection dwindles until both partners' separate identities are reestablished. This diminishment of erotic connection also can happen in a relationship that becomes ritualized, where doing things "as we do them" becomes more important than being open to what else might become possible as each person continues to mature and grow.

Some sex counselors now work with couples in such situations. In *Mating in Captivity: Unlocking Erotic Intelligence,* psychologist Esther Perel discusses how she works with couples to achieve the goal of establishing and maintaining a long-term, passionate sexual relationship. The challenge of sustaining the excitement in long-term relationships, she notes, is that couples want security and comfort, on the one hand, *and* freedom, independence, and lustful sexuality on the other.

The part of each of us that wants security can begin to manipulate or directly control our partners to keep them from straying or even just having too many divergent opinions and behaviors. The part of any partner that wants freedom will be put off by that and want to escape, and so will pull away. To avoid such an outcome, Perel advocates working to see your partner as actually separate, with the right to, and the capacity to, leave you for someone else. Regarding your partner as an ongoing mystery unfolding to you can preserve the magic and surprise in partnered sex and love while balancing safety and freedom.[3] You then can work to integrate safety with autonomy by making it safe for you and your partner to be fully yourselves sexually, even if that is at odds with how you seem in other parts of your life. Such interdependent intimacy can deepen when it allows for an ongoing exploration of who you are and what each of you wants both physically and emotionally.

Establishing such an intimate bond with a partner is a baseline from which it becomes easier and easier to be authentically you in other relationships where different levels of intimacy are appropriate. Plus, the entrainment that develops from communicating through physical touch can unleash more open verbal communication between partners and in some cases allow men and women to become androgynous in their communication styles. Research on gender styles suggests that on average, men's communication styles reinforce their independence, while women's emphasize bonding. Deborah Tannen concluded that "for most women, the language of conversation is primarily a language of rapport: a way of establishing connections and negotiating relationships. . . . For most men, talk is primarily a means to preserve independence and negotiate and maintain status in a hierarchical social order."[4]

Such groundbreaking work, in alerting us to important differences, necessarily oversimplifies the great variation among individuals and how much overlap there is between male and female styles. Yet love stories, in life as well as literature, show how both men and women utilize the language of rapport in romance, courting, and sexuality. This is balanced by establishing personal boundaries and independence through lovers' quarrels and why it is so important that their resolution, in making up, maintains the couple's bond while facing the divergence of belief, opinion, or lifestyle preference, rather than sweeping differences under the rug.

The next lesson follows up with the theme of balancing being true to oneself while also being in relationship, exploring it in the pivotal time of adolescence as reflected in a fine example of teen fiction.

APPLICATION EXERCISE:

Learning to Recognize the Call of Eros

Sitting comfortably, begin by breathing into your heart and out through your solar plexus, thinking back over the call of eros in your life in romantic love, particularly, or in other loves that have had that kind of charge for you, and what you have learned through increasing your powers of discernment. Pick one memory that calls to you now. Then breathe in memories of the joys it gave you and breathe out its pains, letting them go. Looking back, notice what this experience taught you about being true to yourself (or not) in relationship and how you grew as a result of it. Continue until this feels done, and then allow yourself to become aware of where eros is calling you now. As you breathe in, imagine scenes of your desires being fulfilled, and then breathe out the fears that would get in the way of following them. Repeat until you feel that the process is complete.

Persephone Lesson Two:

Claiming Your Love Rights

L EARNING TO STAY TRUE TO YOURSELF while being in relationship with others is a lifelong process, but adolescence is
a pivotal time, so as you read this lesson you might want to
remember when, whether long ago or only yesterday, you were differentiating from your parents, discovering who you were separate from
them and their expectations, beginning to feel sexually attracted to
potential boyfriends or girlfriends, and imagining the kind of life you
wanted to have as an adult. The Zeus and Demeter parts of society
often positively reinforce your finding out what you are good at and
what you care about. But when Persephone calls you, the response
from others may not be so positive, since they may not understand the
changes you are going through.

Let's turn to modern young-adult fiction for an analogous contemporary guess at how Persephone's story could play out in teenage girls
today, judging by books that they choose to read for pleasure.[1] In the
Abandon series (*Abandon, Underworld, Awaken*), Meg Cabot, a bestselling author of teen novels, updates the myth of Persephone and Hades
as a love story between Pierce Oliviera, a seventeen-year-old girl living
on an island in Florida, and John Hayden, who looks her age but is

about two hundred years old. John stays alive because he has taken on responsibility for the regional holding area for the Underworld, which now has retail outlets or inlets because death has become a growth industry as the human population has increased exponentially.

Pierce had met John in a cemetery when she was a little girl and liked him, so when, at fifteen, she drowns in her parents' pool and wakes up in the Underworld, she is not scared of him. How did she drown? She fell in while trying to rescue a bird that looked like it was drowning. It was winter, she was wearing a heavy coat and scarf, and she got caught in the worn-out pool cover and trapped at the bottom of the pool. When she gets to the Underworld, very scary-looking men, on horseback and carrying whips, are pushing people into one of two lines, awaiting the boats to take them to their final destination. In one line people are cursing and fighting with one another, and in the other they are waiting patiently, but all seem to be bewildered and frightened. Pierce does not even notice how terrifying John seems to the others. She goes up to him and complains that the people in line are confused, since there are no signs telling them what to do, the guards are mean, and people are cold, hungry, and thirsty and need to be better cared for.

When John's horse rears and he falls, Pierce immediately expresses concern about his well-being. John is captivated by this beautiful young woman who shows more care for others than he has seen in other recently dead people (as he later mentions when declaring his love for her). When she says that she is cold, too, he gives her the choice to get on one of the boats or go with him to a warm place. She chooses the latter, but does not expect that he would take her to his room. Once there, Pierce is nervous about the bed in the room and what his expectations might be, and when he explains that she is dead, she gets really frightened and runs away.

She wakes up back in the Upperworld, where she is told she has had a near-death experience. When everyone starts asking her if she saw a light and if her loved ones were welcoming her (based on popular descriptions of near-death experiences), she quickly figures out that it is best to keep quiet about what actually happened, especially when her attempts to share what she experienced are met with condescending lectures about how it was just a hallucination.

Of course, John, this manager of the local Underworld, is a rather over-the-top wild and mysterious Hades-type dark man, who shows up to rescue her in many difficult circumstances but unfortunately does so with more force than might have been wise. He comes from a very dysfunctional family, with a cruel and unscrupulous father, a ship captain who threw John and his shipmates overboard when they questioned his plan to risk the lives of his crew to make a few extra bucks. John drowned and woke up in an empty Underworld with the dead starting to appear and realized that he had to run it. Thus, he is wildly unsocialized and has a tendency to overreact.

Because he is visible only as a shadow in the video-surveillance cameras, Pierce takes the rap for what he does to those trying to harm her, so she gets kicked out of one school after another. As a result, she ends up in Pathways, the program for troubled teens in the local high school, with well-meaning counselors who have no clue what is really going on with her. And she knows that if she told them, they just would think she was insane and she would be put in a mental hospital, which would be an even worse fate.

Pathways is in the D-wing of the school along with all the other kids that the principal thinks might be bad influences on the good kids, and those in it are looked down upon by the smart and seemingly well-adjusted teens in the A-wing. However, the lives of the A-wing kids appear to revolve only around a rather meaningless competition between the juniors and seniors about who can get away with breaking school rules by building and publicly burning a casket. This tradition is related to the fact that in the distant past, many people on this island died without a burial, one of whom was John, the sadness of which does not seem to touch the shallow A-wingers.

Pierce is too distracted by her near-death experience, suddenly finding that people are trying to kill her for reasons she doesn't understand, and being in love with the most inappropriate person possible, to focus on her studies. Yet she is highly aware of the educational class system that is not tracking her to a desirable life in an economy with increasing gaps between the haves and have-nots (although her dad is rich, which would give her a cushion), and she is fairly alienated from her schoolmates.

Loneliness often is a theme of teenage fiction because adolescents so strongly identify with it. We can see this portrayed profoundly in John Green's bestselling teen novel *The Fault in Our Stars* (made into a film in 2014), about two adolescents, Hazel Grace Lancaster and Augustus Waters, both dying from cancer and feeling isolated and lonely because their experiences are so different from those of their peers. Hazel, particularly, also is very protective of her parents, who are supportive, but their grief is so understandably palpable that she worries more about them than herself. Hazel and Augustus's budding relationship brings them comfort.

Finding that special friend who can understand what you are going through in the teen years is very important, and Pierce finds this with John. While initially he is frightening to her, she soon realizes that he is even more isolated than she is, and he begins to soften as he feels less alone. Pierce disengages from life around her, but that only makes things worse. When she decides to reengage, as her mother urges her to do, she becomes aware that Internet bullying, which has so wounded her, is happening to other kids in her school, and that a predatory male teacher likely has caused the suicide of a sweet young classmate. She has good reason not to tell her loving mother that the lord of the Underworld keeps showing up in her life, but she also keeps all these other, more typically high school traumas secret too, mainly to protect her mother from worrying or out of a belief that she would not understand.

Think for a moment about your own experience of adolescence, when and why you felt lonely, what you kept secret from your parents or friends, and how eager you were to learn about the larger world out there, even parts of it that your parents wanted to protect you from. Adolescence always has been a time when children differentiate from their parents, often by rebelling and engaging in behavior that their parents would object to. That is where a wild and inappropriate boy-friend sometimes comes in.

When I was sixteen and still quite innocent, I not only wished for a boyfriend, I prayed to have one. The next day, a handsome stranger, several years older than me, arrived in his lavender convertible to a block party on my very street, and I took this as a sign that my prayer had been answered. I started dating him, but eventually found that I'd gotten into a relationship a bit over my head, especially when he became abusive. But at least at that critical juncture, I did not feel so lonely. I had him.

Looking back on that time, I've realized that much more was going on in my life that drove me into the arms of a local wild man. My parents had been taking me to Billy Graham revival meetings as a young teen, where the association of sex with sin put terror into my heart, causing me to go forward to be saved at every meeting—even though my level of sin had only been lust in my heart. (This feels humorous to me now, but it certainly did not feel that way then.) As logic kicked in, I wondered why a loving God would punish unbelief, or being sexual when you were not supposed to be, with eternal torment. In addition to this existential crisis, I was feeling lost in a new and very big school, and I had not yet made new friends. My father was preoccupied, having just lost a business, and my mother had gone back to teaching and was scared that we might not have enough money to get by. I assumed that my concerns would just upset them, and they had enough on their minds already. Besides, since I was a teenage girl, the solution seemed simple: a boyfriend.

Dating him introduced me to a societal "underworld" very different from my caring, straitlaced family. He came from a very different home, where his mother had married an oil-field worker who came home exhausted every day and dealt with it by drinking all evening. Putting things together, in retrospect, I suspect that this man was abusive to his wife and his stepson. I knew that my boyfriend's mother felt she could not leave his stepfather, since she had no education or skills that would allow her to support herself and her son. And looking back, I now realize that my brief experience with abuse motivated my interest in women's issues, because I knew I was one of the lucky ones with enough self-esteem and options to get out of the relationship quickly.

Pierce comes from a more modern, progressive family than I did, so she does not seem to have issues about sexuality being sinful, although she does have theological questions and her own existential crisis. The dominant cultural story she is told is that when you die, you see a light and loved ones who have passed on, but that is not her experience. Having the grown-ups believing stories she knows are not true undercuts her ability to trust that they can help her. Like Zeus, her father is too preoccupied with his business to track what is going on with her (like not noticing that she has drowned), and like Demeter, her mother is kind and loving but would like Pierce to remain more innocent than is reasonable, given what she has gone through.

Even though she feels as though she has no idea who she is, Pierce is clear that she will not consummate her relationship with John until it feels right to her. This does not happen until the end of the second book in the series, and by that time she is at least seventeen, which in the modern world demonstrates restraint on her part. Her concern is less about sex itself than what it means to commit to someone so that you are living their fate as well as your own, an issue that remains very current in a world where women have their own careers that do not always match up with their partners': one of you may need to live in Houston and the other in Sydney, Australia, to do the work that is your calling, or one may yearn to live in New York City while the other dreams of a quiet rural life.

Of course, if the partner's career is in the Underworld, that really would be a challenge. Before deciding whether she wants to be with John, Pierce needs to know more about what her fate is and whether she can live her true calling in the Underworld. Analogous situations today? I've actually heard people tell me that they feel as though their workplaces were populated by the walking dead, who have lost their joy of living in doing jobs or working in a system that kills their spirits. And many divorces occur in a no-fault way when partners find their fates to be incompatible, though that does not remove the pain of such a loss of someone they love. Only when eros and discernment are balanced can such outcomes be avoided or escapes from them be managed gracefully.

Since Pierce is yet neither old enough nor wise enough to know how to use her inner eros GPS, Cabot adds a talisman to the story. John gives Pierce a necklace with a large inlaid diamond that was crafted by Hades and given to Persephone to protect her from the Furies. The diamond turns different colors when she is with people, alerting her about whom to trust and who means to harm her, as well as when she is moving in the direction of her true fate and when she is not. She later learns that the necklace also can free people from being possessed by evil archetypes like the Furies, who are especially malevolent when in the body of a weak person, just as the negative side of archetypes can possess you if you are not anchored in a core sense of who you are, a code of ethics, and a caring heart.

From the time she died at fifteen to when she is seventeen, Pierce sees John primarily when he is rescuing her. However, eventually John kidnaps her in order to save her life, and she recognizes that, at least for a while, the Underworld is the safest place for her. Therefore, she works to improve conditions for all who pass through, so that the experience of dying is not so terrifying and "life" there is pleasant. Pierce focuses on bringing tea to people who have not yet figured out that they are dead, making them feel more comfortable, and making the environment more attractive. A contemporary analogy would be a young girl who moves in with her boyfriend and finds out that his relatives do not get along and his apartment is dreary, so she taps into her aesthetic sense and social intelligence and gets busy renovating and providing tea to help everyone calm down. Pierce begins to notice that sometimes when she just imagines what she needs, the Fates provide it (as they often do when we are on our right path, but without the magic immediacy Pierce experiences as she wishes for things and they immediately appear).

And, as a typical teenager in love, Pierce wants to be with John, even if he is a handful. So she also works, invoking the time-honored Beauty and the Beast archetypal process, to take the edge off some of his wildness, and he begins to act with increasing gentleness and restraint. Given his greater physical strength and magical ability, he moves from exercising his powers to try to control her to using them to protect her and help her get what she wants, which is the classic challenge for males of our species. Realizing that she seems to have a natural aptitude for improving the Underworld, and having upgraded its quality of "life," Pierce decides that she would be a good codirector. I know that when, especially in my consulting work, I sometimes find myself in an organization that is a mess, my first impulse is to think, "Why me?" but my second is to realize, oh, this is my calling to use my talents to create a more human-friendly, well-run environment.

The Abandon trilogy's equivalent to the Eleusinian famine is present in two major themes that widen Pierce's and the reader's focus from personal to larger societal issues: environmental destruction and a failure to honor the dead and death. The setting for Pierce's story is a fictional island in the Florida Keys called Isla Huesos. Pierce's father, the Zeus character, is a wealthy CEO of a global company that supplies products and services to the oil-and-gas industry as well as the military, and the island's rich ecology has been destroyed by a Gulf oil spill. By the third book in the series, a major hurricane is hitting the island, and the seawall that has been built to protect it against such a natural catastrophe is crumbling because of faulty construction (likely a reference to why Hurricane Katrina caused such catastrophic damage in New Orleans).

Pierce's mother, a Demeter character, is an environmentalist who moved to the island to further her research on roseate spoonbills, which are threatened with extinction. She divorced Pierce's father because of his obliviousness to how the way he makes a living makes him complicit in the destruction of wildlife and to his daughter drowning while he was on the phone. She is a good, caring mother who naturally wants her daughter to be able to resume having a normal adolescence but has no idea how impossible that is for her now.

Pierce begins to suspect that the island she lives on is cursed. *Isla Huesos* means the "isle of bones." It got this name because the hurricane of 1846 flooded cemeteries and coffins and killed many people, leaving the bones of the dead spread out all over the island. Merchant ships also had traveled regularly through turbulent waters to what was then a port, and many sank, killing the sailors, whose bones also washed up onto the land. People on the island, called wreckers, would dive down and recover the valuables from these boats and sell them.

Over time, these bones have been cleared away, but construction of a luxury resort is occurring over a site that had been a bird refuge and also was a Native American burial ground, where waves from the ocean have washed away soil, leaving the bones uncovered. All these bones could have been removed and reburied in some place that honored those who had died, but they had not been because of the greed of the owner of the resort, who originally made his money out of cargo from wrecked ships and then from the drug trade. These extreme situations are symptomatic of a society in denial about death, so the thanatos archetype is in shadow but beginning to demand attention—bones popping up, natural disasters, and all sorts of apparently good citizens doing destructive things to others.

Enter the Furies, which are figures from ancient Greek mythology that do not usually appear in the Eleusinian rites but that possess various characters in the book who try to kill Pierce, to get her back dead in the Underworld, where she belongs, and/or partly to hurt John. One of these is her grandmother, who is scheming to kill Pierce in order to destroy John and the Underworld with the purpose of returning life back to the way she thought it should be (likely with more conventional Christian ideas, but her version of Christianity is cleansing the world of what she views as sin and sinners). When Pierce discovers that many of the A-wingers are running a drug cartel, they come after her, too.

To figure out what to do, and to differentiate from others, Pierce has to develop her powers of discernment on many fronts. She needs to recognize that John is basically a good person, even though he has done bad things; her grandmother is a murderer, even though she looks like a nice little old lady; the supposed "good kids" actually are dangerous, while some of the "troubled" teens are very trustworthy; the island is under siege by the Furies; and the Underworld can become a nice place with a bit of fixing up. In this process, she learns to trust her own perceptions in the face of universal disbelief or disapproval; gains clarity that being the queen of the Underworld is her calling, as is being married (eventually, when she is ready) to John; and establishes her own moral code.

At the crisis point of the third book, the Upperworld is being torn apart by a hurricane, while the Underworld is overloaded with the dead and is dreadfully hot. The boats that are supposed to take the dead to their final destination are first late, and then destroyed, while the dead are beginning to riot and ravens are flocking in, looking menacing. Pierce hears many conflicting narratives about what is causing this, some of which blame her or John. However, with a few friends and members of the newly dead who leave the Underworld to help them, Pierce and John save people in the Upperworld from the hurricane and liberate the island from its ancient curse.

In solving the mystery of what is causing the immediate crisis, Pierce determines that the problem is the desecration of the Native American burial ground, which her mother reports to the authorities. Several of the worst perpetrators, including some members of the family that owns the resort over the former wildlife sanctuary and the burial ground, end up behind bars because of Pierce's efforts. In the process, she also resolves her metaphysical crisis by hearing a story that explains all this turmoil that makes sense to her. In it, the Furies (who are the agents of evil) and the Fates (who are on the side of good) each strive to influence events. Breakdown happens when the balance of power shifts in favor of the Furies. While many people are neutral in this tug-of-war, some people devote their lives to good and some to evil. The more who choose to actively do good, the healthier the Upperworld and Underworld become.

The happy ending of this story for the larger community occurs when, through Pierce's efforts, the balance of power goes to the Fates and those humans who devote their efforts to doing good, who now include Pierce, John, and, by this time, the growing group of their supporters, all of whom have committed to the path of making the world a better place. In doing so, they are a bit like the Buddhist ideal of the bodhisattvas—enlightened people who devote their lives to helping others and are willing to sacrifice to do so, even if it means coming back life after life to endure its sufferings. Despite not hav-

CLAIMING YOUR LOVE RIGHTS

ing achieved the Buddhist idea of enlightenment, Pierce, John, and their friends give up their right to die in order to do the work they were called to do as long as that is required (though instead of living multiple lives like the bodhisattvas, they stay alive for that whole time, hence John's being 200 years old).

The Abandon series is a fun and funny story about a plucky heroine, yet the impact of environmental devastation and other social problems on young people today is no joke. Colleagues who are depth psychologists and work with adolescents tell me that many are having end-of-the-world dreams and feel anxious about the threats to their futures from climate change and nuclear proliferation. They also are mistrustful of the older generation, whom they see as having created such problems, as well as the economic fluctuations that make their financial futures uncertain. Often, they do not turn to them when they are in trouble, since the young people see their parents struggling with their own lives and think they have enough on their plates. These psychologists and others are concerned about teen depression and suicide.

Jungian analyst Marion Woodman (in *Dreams: Language of the Soul*) identifies an emergent dream pattern where images of the Black Madonna or other dark goddess figures appear to adolescents, as well as to other people of all ages, as images that point the way to hope and healing.[2] Pierce does not report seeing a dark goddess in her dream, but she does learn from her dreams, and she finds comfort and sustenance in Persephone's mythic story, which gives her permission to follow her own path and find her own happy ending. Because Pierce read the story of Demeter and Persephone in school, she knows that Persephone managed to spend some of her time in the Upperworld and some in the Underworld. The mythic story empowers her to negotiate a similar arrangement (only with John accompanying her), so they also have a condo in town and can split their time between it and the Underworld.

I've chosen this young-adult series to discuss in detail because it is explicitly patterned after Persephone's mythic story, and it reveals many aspects of how Persephone's story could play out in contemporary times. Because it weaves together serious issues teens face in a humorous way, it illustrates the combination of wisdom and lightness that is the gift of the dark goddess in any of us. Most of us, especially when we are teens, have trouble holding difficulties lightly enough that we are not paralyzed by them, but reading these books keeps the reader laughing, even in the face of death. The scenes in the Underworld depict people who do not understand that they are dead—one with an ax still in her head, another with a bullet hole in his chest, and another dripping wet from drowning, as Pierce was when she first arrived. Some of the worst characters begin to make ridiculous excuses for themselves when directed to the line for the bad apples, and one of the sweetest people in the good person line keeps talking about the importance of putting Jesus first, apparently oblivious to the fact that she is standing in a place that embodies a pagan view of the Underworld from an old myth. (She later ends up postponing going to Heaven, where she believes the boat will take her, to join Pierce and John in doing good in the Upperworld and Underworld for as long as her efforts can be useful, so she is a positive example in the book of living the Christian injunction to love her neighbors.)

Much of the wry humor in the book comes from Pierce's typical adolescent wish to fit in at a new school and get along with her family, while having a lord of the Underworld show up all the time and being taken aback by her attraction to him. At first, her family does not know about him, but eventually there is a hilarious scene where she is introducing him to her parents, who assume at first that she must have lost her mind when she explains that he is a death deity who has been saving their island community from the wrath of the Furies. It becomes even funnier when John sends lightning down that burns their carpet and her entrepreneurial father begins to get excited because clearly there is money to be made with this young man. This humor carries with it the lightness associated with Persephone as an archetype and provides the reader with the experience of laughing at difficult life situations, since many of those in the book are exaggerated versions of possible everyday happenings.

To explore how the call of eros has a way of taking many people in just the direction that would stretch their parents, imagine the following scenarios of young adults talking to their parents as they might be staged in a serious drama or in a lighthearted comedy: to conservative religious parents, "I'm gay"; to progressive parents, "I've joined the Tea Party"; to parents who are academics, "I'm not going to college"; to Orthodox Jewish parents, "I'm marrying a Christian"; to good-cause parents, "I'm going to work on Wall Street." I'm sure you could flesh out this list with many more. If your own situation would be on your list, then remember that if eros calls you to do something that would shock your family, eventually you, and they, may remember your exchange as a shared funny story. One person's abduction into a new calling often motivates others to question their biases in order to save the relationship. And in the meantime, you can remember how much more shocking it would be if you announced that your boyfriend is a death deity, saving your city from Furies, who does not kill people—but in emergencies does maim them!

There may have been, or there may be, times when you have followed a calling that required you to face so much resistance, or even suffering, that you lost faith in eros. That is when you most need the next lesson.

APPLICATION EXERCISE:
Taking Control of My Life

Take stock of how good you are at standing firm in your own identity based on the following key factors: trusting your own perceptions when others doubt them; standing up for your beliefs; loving whom you actually love rather than who would look good at your side; staying in connected relationship without caving in when conflicts arise; having a clear moral code and following it consistently; and overall, trusting eros as your guidance system. How different are your current answers from what you think might have been the case when you were seventeen?

Persephone Lesson Three:

Doing Life a Simpler Way

THE SECOND MOST IMPORTANT Persephone lesson is knowing what to do if you have lost your connection to eros or have never had one. Persephone naturally would be taken aback when she is erotically drawn to a flower and then ends up abducted and in the Underworld. It would be understandable that she then would distrust eros for a time. In order to achieve her happy ending, she needed to reconnect with her erotic guidance system. The following myth describes how she and you can do so. The surprising thing is that the steps to achieving this sound difficult but actually can be quite easy to take, as you will see in Psyche's story, which serves as the missing piece of what Persephone did in the Underworld.

The myth of Psyche and Eros is not subtle in letting us know that it is an allegory. *Psyche* literally means "a person's psychology." Eros (Cupid is the equivalent Roman god) is the son of Aphrodite and traditionally is depicted shooting arrows to determine who and what people will love—so he is an embodiment of eros. Thus, it is clear that although the frame is a love story, it really is about Psyche's consciousness and its relationship to her own erotic guidance system. The first known written version of the Eros and Psyche story comes from the Latin novel *Metamorphoses,* by Apuleius,[1] in the second century CE, but evidence in Greek art from as early as the fourth century BCE suggests that it had been known long before. There are ample internal references in the text of the Psyche and Eros myth to conclude that it was part of the larger Eleusinian lore, perhaps even taught in the rites or told as a follow-up that focused very specifically on how people could replicate Persephone's understanding of eros, fate, choice, and destiny to realize a happy resolution to their situations. Or perhaps it feels like the missing piece because it is a complementary element in Greek mythology that helps fill in the blanks that might have been assumed in Eleusinian Mysteries times.

Allusions in the Eros and Psyche story underscore parallels with the myth of Demeter and Persephone: First, Psyche's visits to Demeter and to Persephone serve as bookends at the start and near the end of her story, highlighting their connection with the Eleusinian rites. Second, Psyche's father gives her in marriage to a god, without asking what she wanted or even letting her know what was in store for her, just as Persephone's father did. Third, a god, Eros, who will be the love of her life, abducts Psyche, and eventually they work things out and are happily married. And finally, just as Persephone gives birth to Dionysus, the god of joy, Psyche gives birth to Hedone, the goddess of pleasure and enjoyment. (The negative association of Hedone with hedonism came later through Roman indulgence but was not there in a Greek context.)

Early in her story, Psyche is tied up in the forest at the behest of her father, who agreed to sacrifice her to the gods. Left alone in total darkness, she expects to be devoured by some wild animal. However, Eros carefully unties her and carries her off (again, the father decides, the lover abducts, and the heroine becomes his wife). Aphrodite, who has engineered this whole relationship, has set the ground rules, commanding that Psyche must never look at Eros, so he visits her only in the dark of night. Because this story is an allegory, we can make the connection that most of us start out in the dark about the nature of our souls or the calls we receive from it.

When Psyche's sisters come to visit her in Eros's richly appointed home, they suggest caution: Her husband could be a monster, they worry, which would explain why Psyche is forbidden to see him. Perhaps someday he will try to kill her, so they tell her that it would be wise to keep a dagger for protection and a lantern to see by, and to take a peek at him when he is lying beside her in total darkness. This plot element typically is interpreted as the sisters being jealous of Psyche and wanting to take away her happiness, but at the metaphorical level, it makes more sense if we see the two sisters as the part of each of us that tends to doubt our own erotic impulses, fearing that they will get us in trouble.

When Psyche follows their advice, a drop of oil from the lantern falls upon Eros, and he wakes up and flees from her. However, this one fleeting look shows Psyche how incredibly handsome he is. Now she is totally in love with the husband that she fears she never will see again. Eros's good looks suggest that Psyche is getting a hint of the beauty of her own soul, which Quakers would call the divine spark within, and beginning to love herself. Psyche catches a glimpse of Eros just as we may glimpse our deeper essence, but then he runs off, as our souls often do if we meet their guidance with skepticism.

You may have experienced a time when your own doubts and fears, or those of others, encouraged you to ignore the urgings of your wiser inner self. I know that in my life, there have been times (one when my dark-man boyfriend got abusive, another when Doug died, and others with career setbacks) that I stopped trusting that call from my soul. When we doubt one of our inner capacities, it usually disappears for a good while, until we are ready to trust again. Psyche's story, after Eros has fled, helps us know how to get eros back as our inner GPS, so that we can make decisions that satisfy our deeper selves. To accomplish this, we need to increase our level of discernment about what are calls from our souls and what are not, and we need to recognize that eros often tempts us with what we think we want and then gives us what we need in order to grow.

Psyche is desperate to find Eros, but initially no one will help her. The other gods and mortals fear Aphrodite, which sounds reminiscent of why hardly anyone dared to help Demeter (their fear of Zeus). Psyche appeals to Demeter, who knows all about being abandoned to her own resources in a tough situation. To her surprise, Psyche finds that the temple of the goddess is in disarray (which may suggest that this is during Demeter's sit-down strike in Eleusis), so she cleans it up. It always is good to begin a challenge by clearing out the mess in our minds and hearts, and doing so is a Persephone strength related to letting go and moving on. In gratitude, Demeter tells Psyche what to do to placate Aphrodite and succeed in her quest to reunite with Eros. In our lives, cleaning out Demeter's temple may mean sorting out the messages from our mothers, letting go of the ones that do not serve us, and keeping the ones that do. Then we are ready for romantic, not just familial, love.

Following Demeter's advice, Psyche appeals to Aphrodite, which requires the ability to be humble and vulnerable. She is begging for help from someone who actually set her up, since it was Aphrodite who inflamed Eros with love for her and convinced her father of the need to sacrifice her. Here, Brené Brown's research on vulnerability is telling. In her book *The Power of Vulnerability: Teachings on Authenticity, Connection, and Courage,* Brown describes how openhearted people who have the courage to show up authentically feel less shame and a greater sense of worthiness and belonging than those who try to fit a mold. Taking the risk to be vulnerable also demonstrates respect for, and trust in, the other person, because it shows that you believe they are strong enough to handle what you are sharing and caring enough to help.

Aphrodite first turns Psyche over to her two handmaidens, Worry and Sadness, who represent the tortures most of us experience when love turns sour or threatens to. When Psyche survives experiencing these horrible feelings, Aphrodite gives her a series of seemingly impossible tasks to fulfill—that is, they are impossible for someone without a developed Persephone side. For the first task, Psyche must separate a large basket of very small seeds from different kinds of grains into separate piles by morning. Psyche begins working as fast as she can, but it is clear that she will not be able to complete this task in time, so she gives up and falls asleep. While she slumbers, ants take over and sort them for her. You may have had an experience where you are torn inside, and you sleep on it and find in the morning that your subconscious, where the soul resides, has sorted everything out. Suddenly, the answer to a seemingly insoluble issue is right there, and you know what to do.

The second challenge from Aphrodite demands that Psyche retrieve some golden fleece from a herd of dangerous rams. The task seems so difficult that Psyche decides to drown herself (which we can see as wanting to give up), but when she comes to the river, a reed suggests that she should wait until the next morning, when the rams go in the water to cool off. Then, she can collect the fleece that has been caught in brambles on the river's edge with no danger to her. Why a reed? Reeds aren't strong, but when they are blown by the wind, they bend, so their flexibility is their gift.

Our conscious minds generally see only what we are focusing on. I once watched a training film (*The Invisible Gorilla*) where the audience was instructed to count the number of times a ball was passed on a basketball court. Afterward, those showing the film asked how many of us saw a gorilla walk through the scene. Few did. They showed the film again and there it was, not hidden at all. To succeed in this second task, Psyche did the equivalent of seeing the gorilla, by using her peripheral awareness to notice the small reed and then the fleece in the brambles, instead of staying fixated on the rams and on her despair.

When there is something we very much want and it seems impossible to achieve, the difficulty may be because we are too focused on the specifics of what we are trying to gain rather than on how we want to feel or be as a result of the achievement. Or we may be oblivious to alternative paths to the ones we are on that might get us where we want to go just as well.

Irene Claremont de Castillejo, in her classic book *Knowing Woman: A Feminine Psychology,* posits that masculine attention is more focused and feminine consciousness is more diffuse, although clearly women today have access to both focused attention and diffuse consciousness. You know what diffuse awareness is if you have seen (or been) a mother in a park, talking to a friend rather intimately, but aware of what her child is doing on the playground, and also alert to the clouds rolling in that might presage a storm. This is not a matter of eyes darting back and forth; rather, it is having soft eyes to take it all in at once as a gestalt. We can imagine that Psyche took in the danger of the rams, the potential release from pain offered by the river, and the new creative option presented by the reed in this way. Neuroscientists have credited this slight female advantage (all these differences are small, with great overlap between the genders) to most women having more connective tissue between the left and right hemispheres of the brain than most men do, although this does not mean that men innately lack a connection to both modes of consciousness.[2]

For Psyche's third task, she must collect a vial of water from a stream that flows out of a cleft on a mountain peak at the mouth of the River Styx, which no mortal has ever reached. In this case, an eagle kindly does it for her. Eagles, of course, have a bird's-eye view of things, as well as the ability to see the detail on the ground (so they can swoop down and catch that mouse), an apt parallel to possessing the cognitive complexity to see the bigger picture. The water flows from the source of the River Styx, which is the river that separates life from death. Symbolically, this requires Psyche to see where life and death intersect and to cultivate the intuitive intelligence that connects the conscious and unconscious realms of the human psyche.

Filling a vial with water from a large river also links with the modern idea of thinking globally and acting locally. It is just a little bit of water she is to get, not a whole bucket. I don't know about you, but my tendency when I'm unhappy is to want to make big changes—move rather than renovate the house, revamp my whole diet rather than make small, doable adjustments, and so on. But generally, key small changes are enough for me to feel rejuvenated. In modern organizational development theory, leaders are encouraged to find the leverage points of small interventions that create change without unforeseen side effects that become too massive to fix. Heroines do not have to save everyone in sight; they just need to do their part.

Psyche completes all these trials easily because she has enough trust in life to receive help when offered. An ego stance can make us grandiose and want to tell a great story about what adversity we faced and what strength and intelligence it took to win the day. Not needing to prove that she is amazing frees Psyche up to benefit from what others offer her. Demeter's connected view of the world also can enhance diffuse awareness and "field sensitivity." When tracking a field, it is possible to notice the river, the mountain, the eagle, and yourself and how they might form a pattern, staying open to how you might collect the water and who or what might help. Most of us, of course, would do this only after our hearts sank in considering what seems to be the impossibility of the task; the capacity to expect a miracle is part of Psyche's wisdom.

When Psyche reports back to Aphrodite, the goddess complains that Psyche did not do any of the tasks by herself—in other words, that she took the easy way out. The point of this lesson is that the only way she could have completed the tasks was by being open to receiving help and by expanding her awareness. I suspect Aphrodite's complaint was merely a ruse to get Psyche to sign on for one more test. Like Persephone in the Mysteries, Psyche must show that she is able to travel to and then return from the Underworld, which means her own inner world. Aphrodite hands Psyche a box and tells her to find Persephone and bring back some of her beauty, so that Aphrodite can add it to her own. Also, she must return the box unopened.

Psyche is prepared to die, thinking that is the only way she can get to the Underworld. Then a voice from a nearby tower (perhaps a metaphor for intuition) provides directions on how a living being can enter this deathly realm. The voice warns her to bring coins to pay Charon, the ferryman, to take her by boat across the black river; to throw honey-barley muffins near the many-headed dog, Cerberus, to divert him, so that she can slip by him; and to avoid being distracted by the many ghosts begging for her attention and help. And most importantly, the voice tells her that once Persephone has given her the box with her beauty, she should be sure to obey Aphrodite: under no circumstances should Psyche open it.

This advice is about the descent into our own inner selves. The black river can symbolize our dark and sad feelings; Cerberus, the parts of our inner world that scare us; our inner ghosts, our unresolved issues; and Persephone, our deeper soul self. The tower's warning tells us that it is desirable to feel our feelings and face our issues and our inner ghosts, but we do not have to do so every time we want to consult our deeper selves. Later, they will still be there. Now, Psyche needs to employ focused, goal-oriented attention and a field-independent aptitude.[3]

Psyche follows this advice successfully and makes her way to Persephone, who responds graciously to the request for some of her beauty. But when Psyche is returning to the Upperworld, she breaks Aphrodite's rules and opens the box. Why? Because she hopes to heighten her beauty to increase the chances that Eros will return. This can be seen as evidence of weakness on her part, but it also can be a sign of how much she wants him to take her back.

When Psyche opens the box and its beauty moves into her, she falls into a deep sleep. In the context of the Demeter and Persephone myth, it is important for us to recognize that the turning point of Psyche's story is when she opens the box and Persephone's transcendent beauty floats up to become part of Psyche. Here, Persephone symbolizes Psyche's deeper self, which is always beautiful and good. When you discover that part of you, it affects how you see yourself and how you act, which allows others to see your value more clearly too.

Why does Psyche fall asleep? Because sleep is the time when dreams come to us. I've often joked that it is nice to have a spiritual practice where all you have to do is nap or sleep the night away. Jung and others describe dreams as letters from the unconscious, telling the conscious mind what it needs to know about situations we are in and what solutions might be available. Attending to our dreams links the unconscious and conscious minds. This jump-starts our creativity; it helps us to recognize new callings and act on new ideas. Sleep is the most powerful time for our brains to realign, as brain plasticity starts working in service of what we are becoming, rather than what we have been told we should be or what we have become resigned to as our fate.

In most spiritual traditions, dreams are regarded as prophetic gifts from the divine. That is why during the initiates' procession to Eleusis, they spent the night at an incubation site, where they slept, hoping for dreams that would be healing or that would provide visions for their lives, for Athens, or for their home communities. A widely known Old Testament example is Joseph's interpretation of Pharaoh's dream of seven fat and seven skinny cows as a sign that Egypt would experience seven years of plenty followed by seven years of scarcity, which allowed Pharaoh to plan for the coming famine and avert mass starvation.

The visit to your own soul is not some esoteric thing outside of normal experiences. It is as ordinary as dreaming or meditating or asking your still, small voice within what to do. Psyche's soul can now speak to her through dreams. Remembering her dreams allows her conscious mind to hear what her soul is saying, and thus be led toward the people and experiences that will most fulfill her. Like Sleeping Beauty, Psyche then is awakened by a kiss, this time from Eros, suggesting that her inner eros guidance system now is working.

Zeus then blesses their marriage and transforms her from a human to a goddess, making her immortal, which for us can mean a greater ability to access and trust kairos time, knowing that part of us lives outside linear time, in the eternal. At the end of this story, Zeus has invited all the other gods and goddesses to the marriage of Psyche and Eros, which is followed by a huge banquet that is described in detail, providing a sense of renewed community and foreshadowing the birth of the goddess of pleasure by the gods' shared enjoyment of this event. And all this is because Psyche took the easy way out at every juncture, though a wise easy way.

So as the missing piece of Persephone's story, what does the myth of Psyche and Eros suggest about what went on for her in the Underworld? At first, we can deduce that being abducted would have caused Kore to distrust herself, especially if she had responded innocently in a flirtatious manner when Hades appeared. Applying lessons from the Psyche story, we can imagine that once Kore got to the Underworld, she needed to slow down and sleep, letting the inner ants sort things out for her (figuring out what to do); collect the golden fleece by discovering what was natural and easy for her to do there (caring for the newly dead and enjoying the company of a very sexy man); and gain some sense of the big picture, like the eagle does (recognizing that her mother would be working to rescue her). Like Psyche, she had to avoid getting so distracted by the needy souls around her that she would forget to think about what she wanted. Since Persephone also is a goddess related to vision incubation, she again would wait for a dream to show her the way, which may be where she got the idea to eat the seeds. Finally, she needed to reaffirm her love for the Upperworld and her mother, as well as her growing love for Hades and the Underworld.

All these strategies are easy to follow; it is just that usually we are not taught to pursue them. The path of eros is different from simple answers that are often taught today, such as the idea that you can have everything you want if you just envision it or think positively or show gratitude—all important practices that I recommend but that are not, by themselves, the entire solution to finding fulfillment. Too often the things we think we want actually distract us from what is best for us. The ancient Greeks knew that our lives are constrained by fate, but that does not have to mean they will be tragic. So we now turn to consider the interaction of fate, choice, and destiny in a heroine's realizing the Eleusinian promise.

~~~~~~

APPLICATION EXERCISE:
Choosing a Simpler Way

Take some time to reflect on situations you were worried about that were resolved easily and effortlessly. What was the role of outside help, your openness to thinking differently and seeking options, serendipity, sudden inspiration, or other factors? Then apply what you have remembered and examples from Psyche's story to a current issue in your life, considering all the factors you can imagine that might make something difficult become much easier than you expected.

Persephone Lesson Four:

Making Choices to
Realize Your Destiny

M AKING WISE DECISIONS can feel difficult. Most of us overload our lives with all sorts of things we think we have to do to make a living, be respected and liked, and so on. That is why learning to listen to the voice of your soul with all the noise that comes from other parts of you and those outside of you is important and the subject of this lesson. The ancient Greeks believed that some things in life are fated, such as Kore being abducted by Hades. While their tragic tradition has Oedipus putting out his eyes and other expressions of serious angst, their comic tradition suggests that there is no sense getting bent out of shape by the many events in life that you did not plan on or want to happen. Better to flow with them. The radical part of the Mysteries includes both Demeter and Persephone refusing to remain victimized by becoming creative, not fatalistic. They do this with different styles, but both make adaptive and game-changing choices as they solve the riddle of how to realize their destinies. Demeter demands that a temple be built for her and creates a famine when she recognizes that she does have the power to get what she wants, in spite of fate. In contrast to her mother's, Persephone's action seems very simple and unpretentious. She just eats some pomegranate seeds. But this seemingly small act leads to huge results: the origin of the seasons and of the Mysteries, and the transformation of the relationship between gods and humans.

Twice in the *Hymn to Demeter*, mortals refer to the gods as yokes around their necks, but these complaints occur before Demeter takes on Zeus and before Persephone is returned to her mother. Through participating in the Mysteries, initiates gained a new experience of being known intimately by the goddesses and therefore loved, cared for, and safe in life and in death. Living without so much fear helped mortals solve the great mysteries that confronted them and aided them in making everyday decisions: "Who is it safe to trust?" "Who needs my help?" "Who will help me?" "What does this relationship require of me, and what not?" "What path to take when?" "How can I make the best of this situation?"

If we explore such questions with Persephone's lightness, it is much easier to discern when we are being real and when we are trying to be a whole lot of things we are not. Sometimes I've taken on so much that an image comes to mind that I bet you have seen too. A small car drives into a circus ring and from it emerges clown after clown until it is unbelievable how they all have jammed into it, and the scene gets funnier and funnier. The image of me as that car, and my multiple responsibilities as clowns, makes me laugh and recognize that it is no wonder I feel so stressed: I'm trying to pack too many things into the life of one mortal person. Laughing helps free up options for getting out of some obligations, while beating myself up for overpromising just wastes valuable time.

To listen to that deeper self requires self-acceptance; otherwise, we will fail to recognize who and what we really want. Put in more psychological language, if we are to connect with the wisdom from our souls, we need to become comfortable with our shadow sides. This is illustrated in another example of women's comic fiction, this time with a heroine in her early thirties. How is it, I wondered, that Janet Evanovich's Stephanie Plum mystery stories still make the bestseller lists when (as of this writing) there already have been twenty-one of them? For one thing, the humor is a kind that produces a feeling of warmth that comes from laughing with, not at, someone who reminds you of yourself. That laughter is the medicine that allows you to accept the potential parts of yourself that do not fit current societal ideals.

This gentle humor often is found in novels that reflect the Persephone archetype, and for that reason, this lesson looks at what the Stephanie Plum novels show us about this archetype today, in a very unpretentious heroine. Stephanie has a dangerous, low-status, low-paying job as a bounty hunter going after local criminals who have missed their court dates. She feels that she never really knows what she is doing and has no future, and she frequently has to be rescued by one of the two men she is attracted to. She dresses in jeans and a T-shirt and hangs out with a spandex-wearing, overweight former hooker, Lula, who is her outrageous sidekick and who is known to threaten criminals by saying that she will sit on them and smash them like a bug. Both comfort themselves with sweets or fast food when they get scared, and they both live in very basic apartments in a working-class neighborhood in New Jersey. Stephanie has no idea how to cook and retreats to her parents' home regularly to get a good meal.

These women are far cries from the current women's-magazine notion of the svelte, put-together, confident, disciplined, ambitious, multitasking, successful female ideal, living in glamorous surroundings and accomplished as a gourmet cook, or in what it takes to manage a family and a high-powered job. When we laugh at or with Stephanie and Lula, it makes it easier to accept the feelings and desires that do not fit this superwoman image. When we like a fictional character and laugh with recognition because we know how he or she feels, we lighten up a little by being reminded that we are just human. This is known as bibliotherapy.

In these novels, the Underworld takes on a modern meaning, as the underworld realm of petty criminals and vicious mobsters. Just as the Underworld's psychological journey takes us into our unconscious selves, including our shadows, Stephanie goes into the shadow of our society, where forbidden human qualities are acted out. Stephanie's equivalent of being abducted is being fired from a job. Confronted by a shortage of decent-paying positions, and because she has no money, she ends up feeling forced to take the only job she can get, as a bounty hunter. In this line of work, she regularly finds herself in situations that threaten her life. There are layers to our individual and cultural shadows: at the surface are the ones we can recognize and laugh at, since they do not truly harm us or others; next are those that just need to be attended to in order to avoid trouble; and then there are the shadows we must recognize so that they can be kept under wraps and not lived out. For example, we all have an inner thief, con artist, murderer, rapist, batterer, and so on, and we are more likely never to act on these not-very-pretty human attributes if we avoid being in denial about having such potentials within us. In that latter category, Stephanie has a gun, but rarely loads it or even carries it with her, since she does not want to end up having shot anyone. On the less extreme end, David and I try not to keep ice cream in our freezer. Best not to be tempted!

The harmless shadow elements and humor of the books also carry forward a very ancient tradition of women-only rites—such as the Thesmophoria, which were separate from the Eleusinian Mysteries but also in honor of Demeter—where they were liberated to behave however they saw fit, without worrying about the judgments of men. In a series of scenes in *Ten Big Ones,* Stephanie and Lula are sent to pick up a criminal who has skipped bail—a woman who has held up a Frito-Lay truck, and whose defense is that she was on a low-carb diet and had her period, and just snapped. The driver sitting in the truck did not stop her, as he says she looked like a woman on the edge, and when his wife looked like that, he knew not to get near her. Lula's first response is to declare that everyone needs Fritos to get through their periods. When Stephanie's gun-toting grandma joins them, she says she had always wanted to do that. This leads to an energetic discussion of whether it would be better to have a naked man feeding you the chips or skip the man and have dip. The tone is self-deprecating, with the women bonding around laughing at shared foibles.

Some shadows we see in ourselves and in others require empathy and compassion. In another of the Stephanie Plum novels, Stephanie is on the trail of a taxidermist who keeps failing to show up at a mandated court appearance. He stuffs his specimens with bombs as a way to slow Stephanie down—and she repeatedly ends up covered with entrails. When she finally catches him, he confesses that he keeps missing his court appearance because his Internet connection—on which his business depends—has gone down. Every day, his cable company promises to come, so he dare not leave, but they never actually show up. This universally frustrating experience, which we all can identify and laugh *with*, inspires Stephanie to arrange for her grandmother to wait at the taxidermist's home, should the cable company deign to arrive. Meanwhile, the taxidermist is able to make his court appearance. Who could help but laugh with recognition at such a classic (though exaggerated) cable-company story?

Stephanie succeeds at bringing in the basically harmless people by using her social intelligence to figure out how to convince them to return for their court dates. She also chases very dangerous criminals in the New Jersey underworld, and in a rather risky but comic form of Persephone's trust in life, she can be scared to death one moment and go right back into an equally perilous situation the next. Yet slowly, she learns to figure out who are the violent, treacherous psychopaths, which criminals are moderately dangerous, and which ones just don't want to go to jail. She also learns how to extricate herself from very touchy situations, developing her Zeus strength, her Demeter caring, and her Persephone discernment and ingenuity.

Many of us start out very innocent as we enter the work world, believing that no one means to hurt anyone else, only to recognize that some people are ruthless and will do us in for their own advantage, or in some cases for amusement, without blinking an eye. And yet, we keep going back in, smarter than before, learning as we go. A great challenge for police, and others, today is that they have not been trained in the discernment necessary to read people, especially those of a different race that they see as "other," and particularly if they have not confronted their own racial stereotypes.

Mystery stories with female sleuths, designed for female readers, often are comic, yet they also build psychological intelligence and foster the knack for reading people and recognizing who really is dangerous and who is only a bit wounded and in a pickle. Because Stephanie can laugh at herself, she can lighten up about other people's vulnerabilities and help them succeed, which is a great ability for leaders to have today. In addition, it is important that she has Lula as her partner, since Lula made a living on the streets and thus understands the underworld and is not shy about sharing with some pride what she learned there. It is Lula who often sees the danger that Stephanie would miss without her.

An additional aspect of the humor in these books is how often, in a time when we as women so value our ability to take care of ourselves, Stephanie is rescued by one of the two men in her life. Stephanie's main boyfriend, Joe Morelli, is a policeman, thus a bit of a Zeus "law and order" figure. Stephanie imagines that someday she will marry him. He is from the same neighborhood, he's comfortably sexy, and she feels safe and relaxed with him and his humble home and lifestyle.

Her other heartthrob is Ranger, who is her prototype for what a bounty hunter should be like, since he seems invulnerable to any harm and always able to take down any criminal anywhere. Stoic and mysterious, Ranger, whom Stephanie lusts after, frequently arrives just in the nick of time to save her when she really needs it. What is both frightening and intriguing about Ranger is how at home he is in the underworld and how well he knows how it works. He is, indeed, a Hades figure; the challenge for Stephanie is figuring out what his role in her story is. His ideal function is less as beau than as role model, even though she suspects that he operates just on the other side of the law.

Stephanie may have been abducted by a recession into a career as a bounty hunter, but it is just right for her. She loves the danger and excitement and living the hero story. The trouble is that as of yet, she is not that good at what she is doing. Ranger serves as her ideal. He always seems to know what to do in every situation and has an amazing ability to gain entry to her apartment or any other building and to bring in even the most terrifying of criminals. Not only that, but he also lives in a pristine apartment, with tasteful furnishings and (as she knows from having hidden out there from time to time) sheets and shower gel to die for. Plus, he has great health habits that, over time, she will need to adopt if she is to remain fit and be more successful in her field. He represents an eros call to live her life in a healthier, more conscious way.

The tension in this situation comes from an inability to sort out the eros one feels for a mentor and that which one feels for a life mate. In the chivalric tradition, a knight would develop his higher qualities and tame his lust by loving a woman who is not available, doing everything he can to improve himself to win her regard. And as long as Stephanie is going with Joe, Ranger respects that relationship except for a little flirting and an occasional stolen kiss.[1] I don't know what Evanovich has in store for Stephanie in novel twenty-two and beyond, but it appears to me that Stephanie's job is to resist giving in to her lust for Ranger so as to motivate her efforts and elevate her desires into their higher forms as she also becomes better and better at her vocation. And he is not a sexual predator. He always respects her boundaries as she sets them.[2]

The point here is to notice what the purpose of eros is in any given situation; because most of us have received no guidance in differentiating among various modes of eros, we often interpret it as sexual when it is not. Eros always needs to be paired with discernment, since we cannot fulfill our deepest desires unless we learn to say no to what distracts us from them while we also learn to recognize eros's different forms and what they are calling us to. For example, eros frequently emerges between people who are working together on a creative project or as part of an innovative work team where that energy serves the work. If it is acted upon sexually, it can derail the project. Similarly, eros can be present in confusing ways with therapists and their clients, mentors and their mentees, and teachers and students. In these cases, the energy is there to serve the growth of the client, mentee, or student, who is projecting his or her potential onto the more fully realized role model. Sometimes eros even emerges in the relationship between spiritual teachers and their students, as it did for Rumi. His love for Shams, who was his inspiration and guide, served as a transitional bridge, so that after Shams died, Rumi channeled his love to the divine. Becoming sexually intimate in such relationships defeats their purposes and also can be devastatingly harmful to those involved, especially the one most vulnerable.

Even eros that is in its nature romantic and sexual can become a trap if it limits your full development as a person. Mythologist Christine Downing, in *The Goddess: Mythological Images of the Feminine*, argues that Kore might have recognized Hades when the earth opened, intuiting that he was her fated husband, and thus not showing too much resistance to going with him or feeling too traumatized by the abduction. Yet by itself, this would just be passive. Claiming her destiny requires her to develop a new, and essential, capacity—the power to choose, not just respond, one that has been made difficult for women through much of human history.

If we interpret the abduction literally, we could imagine that when Hades appeared, Kore may have turned and smiled at him, not expecting him to seize her. The violence of such a kidnapping would be horrifying were it real, and to read the myth in a positive and currently useful light, we might regard it as the aforementioned part of their marriage ritual or as an impetuous act on his part, perhaps from misunderstanding her smile, and that he later regretted, after which Hades wooed Kore, hoping she would stay with him. The shadow hovering over this story is the threat of potentially very damaging traumatic events that occur in some versions of the Persephone myth and in women's actual lives, where the only redemption is to make something better from them.

Fate often does have a heavy hand, helped along by the continued human propensity for violence. Many people to whom it throws an ugly curve make wise choices to do something positive with what happened to them. Examples? People I know well include a veteran with posttraumatic stress disorder (PTSD) who ended up going to school to be a psychotherapist to help others like himself; the daughter of another veteran with serious PTSD who now has a social work degree and helps men like him; and a woman with a learning disability who became a first-rate teacher because of her capacity to understand the challenges her students are grappling with. The brutal potential in an abduction or rape is always there for women, even today. And it is a credit to women (some estimate one in five) who have been raped who find a way not just to heal from the experience, but also to use it to become stronger, smarter, and more determined to not let it keep them from living fully satisfying and heroic lives.

A colleague of mine, as a boy, aspired to be a football player. However, fate intervened in the form of polio, which he contracted six months before the vaccine became available. As an adolescent, he became a trainer for his football team, and as an adult, the organizational development director for a huge active retirement community. His very presence is an invaluable inspiration to employees and retirees alike, challenging each of them to do their absolute best, no matter their circumstances. When they whine, he sometimes finds himself unconsciously adjusting his brace, after which they usually adjust their attitudes and stop complaining. He then can partner with them to find a solution to whatever issue they are raising.

A famous case of great disappointment leading to a clarified calling was that of Jung, who desperately wanted to be an anthropologist. However, he had no money and could not get a scholarship in the field. Instead, he studied medicine, for which he could get financial help, and then integrated his love of anthropology into his practice as a psychiatrist. The result was the creation of the field of Jungian psychology, which gave rise to archetypal psychology, which in turn led to this book.

Stephanie feels abducted by a bad job market, taking what she considers to be a lousy job as a bounty hunter. However, she chooses not just to grow up but also to become a heroine, capable of surviving in very dangerous situations and restoring justice. We face choices all the time between living life as heroines or as passive victims. Her fate is to grow up in a marginally middle-class family in New Jersey with a low-status job as a bounty hunter, but what she does with that, the choices she makes—well, that is what determines her destiny.

So let's now return to Persephone and her analogous challenges, here assuming one of the more cheerful interpretations of the story—that she was attracted to Hades in the meadow or that the abduction was just a ritual, not a kidnapping. Persephone's eating food in the Underworld would have held significance in ancient times, since brides, who often saw their new homes only after the wedding, would signal their acceptance of these living conditions by eating food laid out for them, usually by their mothers-in-law, or their rejection of it by refusing to eat.[3] These practices provide meaningful context for Persephone's actions.[4] Whatever happened between Hades and Persephone, her eating the pomegranate seeds clinched the deal, meaning that she agreed to the marriage.

The symbolism of the pomegranate also speaks volumes. The pomegranate is a traditional symbol of the sweetness of life, of fecundity, and of female reproductive capacities. The pomegranate seeds evoke an image of what eggs within the ovaries might look like, and the fruit's red juices have been likened to blood, potentially in menstruation, the breaking of the hymen, and the birth process. It is possible, given all this, that Persephone's act of courage actually was intended to consummate her relationship with Hades. After all, she does return pregnant. The story of her eating the seeds could just be a way to gain support from Zeus and Demeter in implementing her decision to split her time between mom and husband.

But first let's put ourselves in Persephone's shoes and imagine how she came to her decision. She is torn between two desires. First, she feels a strong pull to rejoin her mother and the beauty of life in the Upperworld, but then she experiences an equally powerful urge to stay with Hades and not abandon her work initiating the newly dead into the Underworld. Each time she thinks of choosing between these two desires, she feels sad. But when she imagines doing both, her whole body and soul expand with happiness. But how to do this? She thinks, "I can't have both," and feels truly despondent.

Suddenly, her diffuse awareness allows her to notice the pomegranate seeds Hades is offering, and a plan unfolds in her mind. In Homer's poem, all this happens after Zeus already has agreed that she could go back to her mother. She remembers the law that says that if you eat anything in the Underworld, you must return. You can see lightbulbs flashing on in her brain: this would be her chance to escape her father's dismissive edict that Hades could have her, as well as her mother's potentially oversolicitous caring (which could ensure her role as a mommy's girl who never grows up). Kore is, after all, still an adolescent, needing to differentiate from her parents.

The Virgin Mary also often is portrayed holding a pomegranate, as are other archetypal virgin and mother figures. In this, we can recognize that its symbolism evokes feminine autonomy: in ancient Greece, describing a woman as a virgin meant that she owned herself and did not mean that she did not have sex. Through the mythic example of eating the seeds, Persephone asserts that a woman does not have to act like property or a dependent.

Persephone's story, clearly meant to educate people, tells us that her mother raised her, her father determined whom she would marry, and it would be expected that Hades would call the shots as her husband in the Underworld, which is his realm. By strategically eating a few seeds, she makes a statement of autonomous choice, radically rejecting the idea that she would be dependent on her husband or mother; instead, she will make her own decisions, in this case choosing a bigger and more complex life than either would have seen for her. This change is so profound that it requires her to take a new name, which is when she officially becomes Persephone instead of Kore.

This part of the story harkens back to the origins of the myth, since Persephone was her name in Crete and being queen of the Underworld was then her vocation. Moreover, it is from the myths of Crete that she is portrayed as giving birth to Dionysus, who, according to mythology expert Richard Tarnas, is almost the same god as Hades, only above ground, so he would have a similarly passionate nature. This equivalency also reinforces the idea that Hades is Dionysus's father, even though Hades is the lord of the Underworld.[5] Dionysus being a symbol of joy can suggest to mortal women that they too can give birth to the life they want to live by making freer choices than are usual for women of their time.

I have seen numerous examples of this archetypal pattern of women making choices to realize an identity that requires them to gain ease in moving between worlds. Grettel, one of my two lovely daughters-in-law, grew up in Costa Rica, where she loved her life. However, when studying for a master's degree in the United States, she fell in love with our son Stephen, who is a gravitational physicist, and finding appropriate work in Costa Rica was not an option for him. Grettel, who was a professional hospital nutritionist and later a health-care researcher, made the decision to leave a country that offered a sense of liveliness, humor, and community for the United States, which, although it has other advantages, is populated by more stressed-out, serious, and isolated folk. I suspect that she felt as though she had been abducted, but she is resourceful and adaptive.

She developed ease in traveling between worlds (as in cultures) through consciously deciding to think of herself as a citizen of both her home country and the United States, living this out through annual visits to Costa Rica and hosting frequent visitors from there (including her mother for long periods of time), as well as through constantly keeping in virtual touch. When she felt torn between wanting to continue to contribute to the greater human good through her work and staying at home with her young children, she resolved the dilemma by getting a job doing online research on services for people with AIDS, mainly from home, for a woman-owned company where she could bring a young child to a meeting, if necessary, without anyone raising an eyebrow. Rather than feeling sorry for what she had lost, as she would have had she seen herself as an unwilling exile, she regarded herself as someone who makes positive choices that expand options in all situations.

Many others are in the situation that I've experienced: being called/abducted to a profession that is so new that no well-traveled career paths—or jobs—exist. Many of us, by necessity, move between different fields of work and study (for me, literature, women's studies, marketing, leadership, etc.) and functional jobs (for me, professor, administrator, consultant) and in that way cross-pollinate these fields and related work settings, like bees or butterflies do for flowers.

I suspect that you, too, can think of instances when falling in love with a person, interest, place, work, or cause abducted you into a challenging situation that provided just what you needed to become the person you are today, and how each of these experiences taught you something. Or perhaps you currently are in one of those natural but disorienting "I don't know what I'm doing" phases, where you can use examples such as these to reaffirm your sense that this is just one episode in an unfolding story that will lead you to realize your full potential.

Persephone's choice to eat the seeds can remind us, when we are juggling many different roles, that we actually do have the power to choose how much energy we put where and what we do when—if we are not driven by unrealistic expectations. It helps always to remind ourselves that we are choosing all of it and can refine these choices at any time. Persephone's act of choice affirms all the roles in her life that are authentic for her—wife of Hades, daughter of Demeter, queen of the Underworld, priestess of the Eleusinian Mysteries, and birth mother to Dionysus (though as we will see, his parentage is complicated). While Demeter and Zeus learned to respect each other's rights, Persephone and Hades go further to develop a partnership that is happy, equal, and liberating, and that is why the rise of her archetypal knowledge is essential to men and women trying to live as equal partners in our changing world.

The content of this book, so far, has been designed to give you the opportunity to choose to awaken the qualities of the Zeus, Demeter, and Persephone archetypes in yourself. Part 2, the Zeus chapters, warned against taking in other people's attitudes unconsciously as an interject (something taken in whole without reflection). However, it also is possible to consciously incorporate the best of what you have learned from others into who you are. The former is like the experience of, say, watching TV and finding at the end of the program that all the chips are gone from the bag. How did this happen? The latter is deciding what and how much to eat and staying conscious while dining.

Persephone's pregnancy suggests that joy is incubating in her, and is soon to be born, as it also can be for you. It is wise to remember that something wonderful is always on the way to happening, but you are more likely to notice when it appears if your worldview allows for this to be true. The section that follows provides you with various approaches to understanding Persephone's way of seeing human life, the earth, the cosmos, and our place in the family of things.

APPLICATION EXERCISE:
Using Choice to Fulfill Your Destiny

Part One: Write a one-paragraph story, with you as the hero-ine, about some situation in your past that seemed to limit you. In the story, identify this limiting event as Fate intervening in your life (what happened to you that you would not have cho-sen). Then describe how you made a choice that transformed that situation into your destiny (meaning, at best, a life right for you, and as a fallback, the life that you chose in the context of your fate). Feel free to throw in Furies, if you sense that you have been under siege from forces undercutting you.

Part Two: Think of some situation in your life now where you feel limited by your Fate, and make a list of at least six choices you could make and what destinies each might lead you to. Feel free to choose a couple of wild-card options that might be dif-ficult for you to imagine doing, as they can free your mind for some new thoughts about what you actually might do.

Persephone Lesson Five:

Experiencing Radical Belonging

MANY OF THE PROBLEMS we face today—from addiction to greed to materialism to disillusionment so huge it leads to suicide—result from people not feeling connected to the world, not being centered in any deeper knowledge of who they are, and not having a sense of larger meaning for their lives.

The dominant cultural story tells us that the universe is lifeless and alien, that it, the earth, and we are a product of random events, and that what happens in your life and mine is equally random, has no intrinsic meaning, and then we die. If that were not enough, many of us assume that having an unhappy childhood dooms us to an unhappy future. No wonder people are depressed or believe that the only purpose of life is to make money. There are many narratives out there that bring individual peace and a sense of purpose—some from dominant religions, some from spreading new age spirituality, some from indigenous traditions, some from transcendental, existential, or humanistic philosophies, and some that people create themselves. Also, advances in psychology and medicine help many people heal from trauma and abuse.

What is shared in this section reflects an ancient Persephone mind-set and its modern equivalents. It is intended to contribute to your thinking about what outlooks and stories can best give you hope and allow you to feel safe and happy and at home in the world. For many of us, our ideas of what the world is like come from our origin stories, meaning the stories we tell about our birth, our family of origin, and how we came to be who we are. Persephone's origin story is quite different from her father's or mother's. She grew up adored by Demeter, and although her father, Zeus, was distant, she may not have minded, since having him for a dad gave her high status. In addition, she would have learned from Demeter the loving and connected values taught in ancient puberty rites. All this was her quite propitious fate, preparing her for her destiny. Your fate may have been more difficult than hers, but any of us can avail ourselves of the quality of her consciousness through substituting stories that support it for ones that do not.

In hunting and gathering times, the archetype that Persephone embodies, in her adult role, would have been the medicine woman. Her functions included facilitating life transitions such as ushering new life into the world; the movement from being a child to becoming an adult; rituals of commitment such as marriage; healing; and passing from this world into the next and potentially coming back again. Along the way, this shamanic figure would need to make friends with some of the monsters in her shadow blocking her path, just as Hushpuppy faced her beasts. Persephone's abduction helps her become who she is meant to be. After exploring what Persephone (and girls of ancient Greece) might have been taught, this lesson will turn to current narratives that update the wisdom of this archetype for our times.

It is believed that the Eleusinian Mysteries originated in these very early hunting and gathering times, since why else would Demeter be credited with teaching people the secrets of agriculture? Most likely the Mysteries themselves began as puberty rites where mothers and daughters would gather to share secrets about "the birds and the bees" as understood in their time. From the era when these rites were in their earliest forms through the later Classical Age, girls would have mated or married soon after puberty, so they naturally might feel nervous about the possibility of leaving their mothers and living with their new husbands, even if in the same community. They might still be in the "sex—that sounds yucky" stage, although beginning to have sexual feelings and being curious about what sex entailed. The idea of giving birth could have been exciting, since becoming a mother would give them status as well as fulfillment—but they also could be terrified of the possibility of dying in childbirth, which was not uncommon.

They would have learned to feel connected to the natural world and to one another as described in part 1, the Demeter section, particularly in the story of Hushpuppy and in the lessons that focus on the reality of mortality. But these puberty rites went deeper than what you would tell a six-year-old, into the heart of pubescent girls' fears. They communicated a belief that sexuality and the reproductive process were deeply spiritual, demonstrating a woman's oneness with the earth as a mother who blesses agricultural abundance and women's sexuality: sexual union, menstruation, giving birth, and nurturing children. Girls may have been taught that their husbands would plant a seed in them, but the metaphor more often used in the Classical period (and that actually was part of the ancient Athenian marriage ceremony) was that in the sex act, the husband would plough in a way that created the condition for the seeds of babies to grow.

After sexual intercourse, it would seem as though nothing was happening for months until the girl's midsection would grow, and eventually she would be gifted with a child. This was seen as analogous to a seed being planted in the earth, and after a long period of waiting, a sprout would appear. Similarly, when you die, you are planted in the ground, and for a long time nothing seems to happen, but then you are reborn (with the forms of this ranging from reincarnation to living on in some kind of afterlife). In these ways, the microcosm of one's own life and biology paralleled the macrocosm of the earth in its processes. From understanding this, girls came to trust that when they died, it would not be the end, just as it was not the end when a plant withered and went to seed, because once that seed was planted, it soon would sprout with new growth.

This view of the interrelatedness of our lives with the processes of the earth is at the root of the Persephone story. As we have seen, scholars traditionally view her story as the embodiment of Demeter's teachings, with Persephone being the seed that goes below the earth and then returns in the spring. She thus reassures mortals that death is not the end for them either, but perhaps even more importantly, that the seasons and patterns of their lives are part of a larger whole where everything is connected and of a piece.

Persephone's story gives life to the philosophical idea that we are a microcosm of the whole that mirrors larger processes, in a manner that can help us feel that we are at home in the universe, on the earth, and wherever we go, as Persephone learned to do. Many scholars believe that a ceremonial *hieros gamos* (or "sacred union") was part of the most secret and sacred moments in the Eleusinian rite, which included an observance in celebration of the marriage between a god and goddess. Although this ritual can be acted out sexually by a priest and priestess, it rarely is (except, perhaps, in fictions such as Dan Brown's popular novel *The Da Vinci Code*).[1] In the Mysteries, it likely was evoked in symbolic form with ritual objects suggestive of a phallus and vagina (which is the underlying symbolism of the chalice and the blade in Arthurian legends and elsewhere).

In the ancient tradition of alchemy and in the thought of C. G. Jung, the hieros gamos represents any union of opposites, and particularly that of the masculine with the feminine. In that way, it symbolizes androgyny and gender partnership as well as marriage and sexual union. Both alchemy and Jungian psychology view the integration of one's masculine and feminine sides as necessary for wholeness and spiritual development.

In the experience of the hieros gamos, sexual union feels as though you both are being taken over by eros, a force that is larger than yourselves, and in this way spiritual. We do not learn the details of how it is that Persephone came to love Hades; we just know that celebration of the hieros gamos in the Mysteries is yet another indication that she did. When a couple mutually commits to one another, sexual union sometimes can be experienced as a spiritual act transcending separateness. If the couple can keep this connection alive, their happiness and bonding fuel their abilities to create a supportive and loving environment for children or other dependents and to contribute to the health of their community. This is especially true if the erotic connection is supported by Demeter's caring and Zeus's organizational skills.

The archetype of the hieros gamos links the mystic relationship of the human to the divine with human sexuality as two analogous expressions of eros. The spiritual poems of Rumi, artfully translated into English by the poet Coleman Barks, are filled with erotic images, and Rumi is now the most popular poet in the United States. The Judeo-Christian Song of Solomon interprets the relationship between God and his people as being like a marriage between a husband and wife, depicting God as courting Israel and wanting to be loved by her. Although without the erotic imagery, the Reverend Donald Miller, in *Storyline 2.0: Finding Your Subplot in God's Story*, construes the Bible as a love story about God's unrequited love for humankind. He then urges readers to acknowledge and show their love for God by living their purpose as a subplot in this larger cosmic love story. Within this metaphor, God's love is realized when human beings return his love, just as in human life, when the person we are in love with loves us back.

The love-story narrative also has moved into scientific thought. In *The Universe Is a Green Dragon*, Brian Swimme, an astronomer and professor at the California Institute of Integral Studies, describes gravity as the cosmic attraction between objects that holds the universe together. It creates galaxies and keeps the earth spinning around the sun and the moon revolving around the earth. It also holds each one of us on the planet. Swimme characterizes this cosmic allurement as the basic binding property of the universe and calls it love. How interesting to re-vision the universe as the stars and the worlds creating loving relationships with one another, and Mother Earth showing her love for us by keeping us on the planet, wrapped in this invisible, loving attractor field.

Swimme goes on to suggest that such an attraction is at the root of human fulfillment: "Each person discovers a field of allurements, the totality of which bears the unique stamp of that person's personality. Destiny unfolds in the pursuit of individual fascinations and interests. . . . By pursuing your allurements, you help bind the universe together. The unity of the world rests on the pursuit of passion." Without this allurement, galaxies would fly apart, the earth would drift off into space, and human community and relationships would break apart.[2] This is not so different from Demeter's web of care, except here the attraction is erotic rather than the caring of the heart.

The concept of the hieros gamos expresses a deep spiritual truth that experiences of union can be like nesting dolls, smaller ones within larger ones. It also can symbolize the union between you and God or the universe; of you with you; of you with your partner; between you and your right work; of your past with your present; and between men and women in partnership in an organization or the world.[3] In all these ways, eros is the key to finding your place in the family of things, helping you to discover where you belong in its whole.

The nesting-doll image of smaller systems within larger ones can be seen as implicit in contemporary scientific views that regard the earth and the cosmos as living systems that have their own kind of intelligence. Some scientists, such as James E. Lovelock (in *Gaia: A New Look at Life on Earth*), have examined contemporary scientific data and concluded that it does not support the prevailing notion that the planet is merely a thing and that changes to and on it happen only through random events. Of course, the idea that the cosmos and the planet have intelligence does not mean that theirs is the same as human intelligence.

Scientist and philosopher Elisabet Sahtouris (in *Gaia: The Human Journey from Chaos to Cosmos*) traces the idea of a living Earth and living cosmos that are becoming so important to the environmental sciences today to ancient goddess cultures such as Çatal Hüyük, located in modern-day Turkey, where archaeological remains are well preserved, down through Crete and then to Greece, lasting in some form into the Classical period, and kept alive by the Eleusinian Mysteries.[4] As you know by now, Greek myths included generations of gods and goddesses. Before the Titans, there was Gaia, who was not viewed as a goddess separate from the earth. She *was* the earth, not just a symbol of it, constantly creating and recreating herself, having given birth to all that we know. There is a huge difference between thinking of Earth as having life on it and it being a living Earth. And the latter is a better match for what many scientists now know about the earth's capacity as a self-regulating system with its own form of intelligence, than the former, more conventional view. In the movie *Avatar*, scientists on the mission discover that the planet Pandora is a living being, and every part of it has access to its intelligence. This plot exposed a mass audience to imagining how such a planet might be and feel to its inhabitants—somewhere, if not here.

Sahtouris traces the development of two major schools of thought about the world in ancient Greece:

> one that all nature, including humans, was alive and *self*-creative, ever making order from disorder; the other that the "real" world could be known only through pure reason, not through direct experience, and was God's geometric creation—permanently mechanical and perfect behind our illusion of its disorder. This mechanical/religious worldview superseded the older one of living nature to become the foundation of the whole Western worldview up to the present.[5]

The ancient Greek philosophers, however, did not experience a dichotomy between reason and their various mystery traditions. Their focus on reason actually emerged from those traditions, as has been demonstrated by Peter Kingsley and others. The split came later.

Continuing, Sahtouris argues that after the move away from the belief in a living Earth came the ideas that humankind is separate and above nature, and God is separate and above all creation. Together, she concludes, these later beliefs form the philosophical basis for actions that have resulted in humankind disrupting the earth's ecosystem, forcing the earth to take steps to recreate balance that now cause unusual weather and that potentially can threaten life as we expect it to be.

Essential to the view of the earth as a living system is the concept that like creates like. The earth creates humankind to fit into its ecosystem, just as it fits into the ecosystem of the universe. Within each of us is a similar ecosystem of other organisms necessary for us to thrive. An example: I started feeling a bit ill not too long ago and learned that a particular food I was ingesting was throwing off the balance of my intestinal bacteria. I would not thrive until these micro-organisms essential to my health did. (Who knew?) Sahtouris credits the philosopher-scientist Arthur Koestler with calling each of the nesting dolls or systems within a system "holons" and the universe of which holons are a part a "holarchy." So I discovered that I have holons within me; I am a holon; and I am part of numerous holonic systems, all of which are part of the holarchy.

Sahtouris claims that this means that "no being in nature can ever be completely independent, although independence calls to every living being, whether it is a cell, a creature, a society, a species, or a whole ecosystem. Every being is part of some larger being, and as such its self-interest must be tempered by the interests of the larger being to which it belongs."[6] Given this, political and philosophical debates about whether individuals should sacrifice their interests to the collective or do whatever they think is in their own interest, hoping that it will all work out, are irrational. Humans have to balance both, as does every other holon.

Sahtouris and others regard this growing scientific awareness of our relationship to our bodies, the earth, and the cosmos as similar to what many ancient scientists and philosophers understood. The mythological image of Gaia continually creating herself in an ongoing dance was not some primitive way of thinking. It was a translation into metaphor of a complex awareness of the actual processes of the earth being created as part of the larger cosmos and then continuing to create itself until it has become the earth as we know it.

Meditating on this, I realized that if this vision is true, then it is possible to overcome alienation by feeling at home on the planet and in the cosmos, not just with our families or close friendship or work groups. We can achieve this to some degree by recognizing that we are characters in a very big story and part of the earth as a living system. We can choose to see ourselves as radically belonging, instead of alienated, because we are part of such a whole. When we feel stuck, we can remember to tap into the power of the imagination to inspire our efforts, since we are a microcosm within a macrocosm that also is always creating. Moreover, the idea of the earth as a living system does not throw out the concept of the divine infusing the whole, although it does not require that this be true either.

Even without quite saying that the earth is a living system, Albert Einstein and quantum physics already have transformed how we see the world working. Chaos, complexity, and systems theories now allow us to understand physical, social, and psychological realities as complex, interrelated systems. Today, organizations, families, and communities all can be viewed as evolving systems, influenced by everyone in a way impossible to track (with infinite causality at work) and self-organized by the interface between the choices each of us makes. Thus, the new themes that govern workplace success, according to Margaret J. Wheatley and Myron Kellner-Rogers, in *A Simpler Way*, are playfulness, creativity, and group self-organization.

Wheatley and Rogers explain that what appears to be chaos in the universe has been found to be always self-organizing; thus, the universe naturally creates order on its own, which allows us to relax a bit, since we do not have to control it all. This reassurance is based on patterns that scientists observe that look like random occurrences. Called fractals, they are self-creating forms that are the archetypes of the natural world. Wheatley and Rogers argue that organizations and other social systems work similarly. Wheatley concludes her seminal book, *Leadership and the New Science*, with this riddle, quoting science writer K. C. Cole, as a great example of this phenomenon: "How do you hold a hundred tons of water in the air with no visible means of support? You build a cloud," of course.[7]

Because some problems solve themselves, we don't have to exert ourselves so strenuously if we notice the archetypal patterns at work in nature, social systems, and our own psychologies. Perhaps someday similar but undiscovered patterns will explain synchronous occurrences—those meaningful coincidences that in some key moments make our lives better. These can be relatively ordinary events, such as meeting the person we will marry, or sitting on the bus next to someone who tells us about a job that turns out to be just right for us, or suddenly seeing a rainbow and feeling our hopes rise. Or they can be major: in the movie *Interstellar,* a wormhole—an opening between galaxies—appears just as humans need it to escape a dying Earth.

The theme of *Interstellar* is similar to that of *Avatar,* discussed in part 3, but draws on actual modern theoretical physics, while *Avatar* is more fantasy based. On the famine-struck Earth of *Interstellar,* the deficit of Demeter is palpable: The mothers of both daughters in the film have died; there is a grandfather but no grandmother; and Mother Earth is dying, which we can see as similar to Demeter's famine, only with this one, it is too far along to be fixed. The planet is becoming uninhabitable because a blight is destroying all the crops in monoculture systems of agriculture without enough diversity for sustainability, giving rise to dust storms that make the air unbreathable. Demeter has not created this disaster, but the absence of her way of thinking and behaving has. The main character's father explains that this happened because the billions of people on the planet all wanted more and more, and a schoolteacher similarly blames it on unspecified human causes.

Moreover, father-daughter partnerships save the day, and it is the two daughters who rescue humanity, one of whom finds a habitable planet while the other solves the problem of how to get humans there, though neither could have done this without their dads. The main character, Cooper, is a former astronaut recruited by what is left of NASA to help find a planet suitable for human life, in the hope that the chief scientist, Dr. Brand, can learn to harness gravity to move large numbers of people from the earth to this new home. Dr. Brand's daughter, Amelia, is a biotechnologist, and she and Cooper, along with several other scientists, are the ones who journey to very strange new worlds—one inhospitable planet after another.

Parallel to Zeus, Psyche, and Persephone's feelings of being abandoned by their fathers, the film's personal drama comes from the strained relationship between Cooper and his daughter, Murphy. Murphy suffers devastating grief and disillusionment, believing that her father has left her on a planet that has turned into a hellish dust bowl, where soon everyone and everything will be dead or dying, to die—potentially saving himself by taking off into space. Yet it is Murphy who has the intuitive qualities that allow her to receive messages through time and space from her dad, and later, as an adult, the scientific brilliance that enables her to understand what is being sent and build on it.

As Cooper travels through space, he is suffering, knowing that he is aging more slowly than those on Earth. He might never get back, and if he does, his daughter and other loved ones may be dead. Yet he persists courageously against great odds because he knows it is his calling to keep searching, and it is the only way to return to those he loves. At the end, he does lovingly reunite with an aged Murphy, who is on her deathbed on a NASA space station to which humanity has been evacuated.

Although *Interstellar* implicitly holds out the hope that humans could relocate, much of the film takes us to planets where human life could never survive, much less flourish. The feeling is dismal, reminding us of the preciousness of our home planet and how unlikely it actually is that we could find a new home elsewhere. A subtheme of the movie shares speculative ideas in theoretical physics that build on the concept that time and gravity are on a continuum. Theoretically, what happens in the movie—finding another planet and moving a large group of people off our dying one—could come about. The challenge is that, as humans, we live in three dimensions. There may well be other dimensions, but we do not experience them. If we did, much could change.

Cooper's ability to communicate with Murphy[8] across time develops because his capacities evolve (and he falls into a black hole), so that he enters another dimension beyond time, gravity, or space, with his persistence fueled by his intense love for his daughter and his desire to get back to her, tempered by his commitment to saving his species. His finding himself in another dimension adds a shamanistic element, since in myth, at least, and perhaps in reality, medicine people go to other realms in the process of healing a person or a community. It also evokes the living-planet, living-cosmos notion of holons within holons in the idea that we might exist within other dimensions as yet unknown to us.

In the film, exploratory vehicles have been sent to twelve planets that may be capable of supporting human life, but Cooper's spaceship does not have enough fuel to get to all of them. The key to finding the right one is the feeling of love that draws Amelia to it because the man she loves went there. She argues that following love may be a rational decision, suggesting that she understands that eros is a guidance system, but Cooper does not initially accept her argument as valid. Yet the planet to which she is drawn turns out to be the only one of the twelve that could possibly sustain human life. By the time she gets there (which she does alone, after Cooper detaches his shuttle from the spaceship so that the spaceship has enough energy to go around the black hole), the explorer she loved and who drew her there has died. Yet it is the power of her love that enables her to find a new home for the humans who have escaped the dying Earth. When she arrives on this planet, she has with her containers full of human eggs, with which she is to populate the planet if no one else can get there. At the end of the movie, Cooper is on his way to join her.

However much we explore new worlds, we all want to find home, in and around us. Psychologist Cynthia Hale argues (in *The Red Place: Transforming Past Traumas Through Relationships*) that even though emotional and physical violence, atrocities, and disasters traumatize and frighten us, healing comes from the foundational pull we all have to find a community where we are valued and loved. Inherent in such communities, and in all of us, is a desire to aid one another, so that, to use her term, we can "holler" one another home through offering acceptance and love.

It is doubtful that Cooper could have persisted through all the difficulties he faced if his daughter's love was not drawing him, or that she could have kept engaged with her scientific work while living on an earth that was becoming unlivable without discovering the messages he was sending through the power of his love for her. And it is difficult to imagine that Persephone could have been so resourceful in the Underworld if she did not trust that Demeter was looking for her. Personally, I would not count on love to help us find another planet fit for human habitation. But our love for one another and the earth, coupled with scientific intelligence, just might help us make the changes needed to restore ecological balance in time, right here, by hollering us home to our better selves.

Choosing to love, in all its manifestations, is one way that we firm up the web of connection between us as people. Imagining ourselves in web-like systems of self-creating life can help us to recognize that what has been called fate may be the product of living in interdependence with all that exists. It is the consequence of all the jockeying that goes on between diverse interests, actions, and desires that eventually affects us as we in turn affect the whole. Sahtouris contrasts a living-system vision of the earth with the deist vision of our American forefathers, who imagined God as being like a clockmaker, creating the universe as a perfect machine. The even more contemporary secular view tells us that humans may be either the only intelligent life in this universe or that such life is a phenomenon so rare, and exists so far away, that we may never find its equivalent. On the other hand, modern entertainment media present us with images of powerful aliens coming to destroy us or as saviors arriving just in time to rescue us from ourselves.

Since there is evidence that Earth is a self-organizing system, we could tell a story of a living Earth and living cosmos, which might leave us feeling less lonely. Sahtouris suggests normalizing our contemporary human predicament by imagining Earth as being like a mother with all sorts of children running here and there and working hard to balance all their interests and keep things in equilibrium. So when our lives are hectic and our multiple roles are running us ragged and others are not doing what we think they should be doing, we feel as if the earth itself, and maybe even the cosmos, shares our conundrum. Yet even then, the interaction of fate and our choices fashions our destinies, because we are not islands unto ourselves but are part of an ever-evolving universe. So what is natural in life is not having everything perfectly under control; rather, it is constantly adjusting to inner and outer information, living life as a self-organizing, ongoing dance. In this spirit, we are about to turn our attention to the god of dance, who represents the culmination of the journey described so far in this book. But before we do this, let's take a moment to return to the role of kairos, or eternal time, which was mentioned briefly in part 2, the Zeus chapter.

Persephone's ability to travel easily between the Upperworld (ruled by linear time) and the Underworld (ruled by kairos), suggests that she knows the secret of how to move from one kind of time to the other. Persephone wisdom is present in kairos moments of insight when you suddenly know what you are to do; in a visitation from a muse that furthers your creativity; or in the feeling of time standing still because what you are doing feels so right and your attention is completely on the now. It also is there when you experience those transcendent mystic moments of oneness where you open up to possibilities, become more receptive, and feel more loved and loving, as well as somehow lighter and less worried.

I once had an experience of the material world suddenly becoming translucent, so that I saw through it into one where all was light, and I felt infused with love. It so transformed my sense of reality that I'll never forget it or be the same. Just thinking of it makes me happy. The humorous part is that I was careening down a four-lane highway at the time, but happily, I managed not to wreck the car. To me, this has become a lived teaching moment that showed me how such instants, when the transcendent peeks into our lives, can transform or wreck them, depending on our ability to stay grounded in material reality when they happen and subsequently as we attempt to integrate their lessons.

Prominent journalist and social activist Barbara Ehrenreich (*Living with a Wild God*), who has a Ph.D. in cellular biology, recently began to share mystical experiences from her youth when she felt "illuminated," but also "shattered," since she did not then and still does not have a belief system that explains them. She confesses that, even now, "the word 'spirituality' creeps me out." Yet particularly in one flash of epiphany, "the whole world came to life, and the difference between myself and everything else dissolved—not in a sweet, loving, New Agey way. That was a world flamed into life, is how I would put it."[9]

Could it be, I wonder, that such moments offer glimpses into another dimension that can provide perspective to all the hustle and bustle of our three-dimensional lives? T. S. Eliot's four-sequence poem, the *Four Quartets*, begins with a reverie about such an experience in a rose garden, where he had a mystic vision that transformed his world-view and also reaffirmed his vocation as a poet. In this lengthy and very mystical and wise poem, he contrasts the lesser world of past and future with such transformational moments, but in the closing passage, he quotes theologian Julian of Norwich, who found comfort in such experiences, concluding that they meant that "all shall be well, and all shall be well and all manner of thing shall be well."

Part of what makes any of us cynical and despairing is the belief that tangible, material reality is all there is. In everyday, practical terms, kairos moments are what can keep you plugged into your soul's deeper knowledge about what you should do, and thus enable you to make choices that are wise because they are on target for your purpose and the call of your future. However, as I learned and as Eliot writes about so beautifully, the insights that come from the eternal through the soul have to be lived out in the body and in linear time. Part 4 explores how the god of dance and joy and embodiment can help you do this.

APPLICATION EXERCISE:
This I Believe

Part One: Write or tell a story about what you believe that provides you with hope and that helps you feel like you belong in this world. Do not be concerned if your story is very different from what was shared in this final lesson from Persephone. What is important is that your story feels right for you. If you find that your description of how you think the world works sinks you into despair or lethargy, then tell a story that would, if you believed it, transform your life in a way that is right for who you are now and who you want to become. You might occasionally, then, act as if this vision were your truth and see what happens.

Part Two: Think back on how Persephone's initial strength comes from a Demeter-infused worldview, later augmented by Zeus's can-do spirit, both of which then are integrated with her understanding of eros as a guidance system, augmented by discernment. This exercise is designed to help you do this too. Think of role models in your life who have embodied Demeter, Zeus, and Persephone's virtues—people who have influenced your thinking and your values, who might be your parents, best friends, mentors, or other role models. Reflect on what each taught you. Then, looking back over your "This I Believe" statement, identify what part of what you have said echoes these influences and what part is uniquely your own. Consider whether you are living out values gained from these role models that are not right for you. If so, take time to reframe them into ones that will support you and your growth. If you have consciously integrated helpful advice and modeling, you might take some time to feel your gratitude for those who have influenced you positively and for your ability to stand on their shoulders while also being true to yourself.

CAPSTONE EXERCISE:

Dialogue with Persephone

In a journal or notebook, dialogue with Persephone where you speak as yourself and the Persephone of your imagination speaks for her. Start by thanking Persephone for how she has enhanced your life, and then ask how she would like to be expressed in you. You can then say what you would like from her, including where you think she would be helpful and where not. Allow a natural conversation to develop where you come to an agreement about Persephone's future role in your life. End on a note of appreciation.

Part Four

~~~~~~

DIONYSUS

Dionysus and the Gift of Joy

IN HIS ROLE AS THE CULMINATION and "fruit" of the Eleusinian Mysteries, Dionysus is known primarily for being the god that brings us joy, and particularly the kind that comes from feeling connected with others, with the body, and with its capacity for spontaneous self-expression. He also is the god of chaos, who helps us dance to the music of life, however discordant it may sound, and feel at home in a world we cannot control, since it is bigger than we are.

I don't know about you, but my life often feels chaotic as I balance multiple roles, doing what is urgent and most necessary while maintenance tasks (like filing or cleaning) get sacrificed. My first impulse is to rely on time management to solve this problem, keeping everything on a calendar and trying to plan ahead, sequencing all my tasks and determining what could be done in what amount of time and where that could happen. But the unexpected always messes up my system. My desire to keep everything under control leads me to feel more and more tense and stressed. Dionysus, however, teaches me to feel connected with this chaotic world—to dance with it rather than fight against it.

Mathematician and philosopher Ralph Abraham (in *Chaos, Gaia, Eros: A Chaos Pioneer Uncovers the Three Great Streams of History*) cites ancient Greek philosophers as identifying the three major forces of the universe: chaos (out of which order emerges), Gaia (our Earth), and eros (the life force and principle of relatedness). Dionysus, the god of chaos, frequently is paired with Apollo, the god of aesthetic order, but in the Mysteries, he would also be the complement to Zeus's focus on social order. Just as Demeter is the goddess who helps people learn to partner with Gaia (the earth herself and the earthiness within us) and Persephone helps us trust eros as a universal force in and beyond us, Dionysus helps us to find joy through playfully engaging with chaos as a larger force in the universe and the everyday force in our lives. To stop fighting chaos, it is helpful to apply the nesting-ball principle of hieros gamos, which allows us to recognize that the chaos we experience in our lives is the same force that inhabits natural and social systems.

Abraham goes on to connect the ancient belief that Gaia and eros emerged from chaos with (1) modern chaos theory in the physical sciences, which holds that apparently chaotic and random behaviors in physical, as well as social, systems frequently form patterns that are nature's archetypes, often called "strange attractors"; (2) the Gaia hypothesis, which tells us that the earth is a self-regulating system with its own form of intelligence; and (3) "erodynamics," which studies the complex symbiosis of human populations with the biosphere. While all these may sound very esoteric, they can help us view our relationship to the universe, the earth, and one another very differently, for at the root of the complexity in our individual and collective lives is an emerging order that contributes to our realizing our destinies. Whether or not you know much about modern physical and social science, you most certainly are aware that you are part of the earth's ecosystem; that you are interdependent with other people economically, politically, and through myriad global companies, nongovernmental organizations, and other organizations and movements; and that you can interact with folks around the world through the Internet and social media if you choose to do so.

However, such advances also cause people to feel more isolated and driven, leading to a hunger for connection with others and for authentic experience. Yet these cravings, which lead some to feel the need for continuous virtual interaction or to seek intense, yet artificial, relationships (reality shows, porn, intimacy with strangers), cannot be satisfied completely through digital means. In addition, ordinary people no longer are content to let a small elite control their lives; they expect to have a say in how they are governed and an equal opportunity to benefit from social, economic, and technological advances. They want to control their own destinies, but their experiences, and the larger natural, social, and economic developments that influence all of us, are being shaped by too many forces to be able to predict what will happen next. And change is quick when an idea can go viral in less than twenty-four hours.

In these ways, Dionysus is a god for our times, particularly since much of the unhappiness in the modern world is caused by the discrepancy between how we now know the universe and our world operate and anachronistic stories about them that determine what we do. For example, we are encouraged to think that we should control everything, when so much is out of our control; to focus on besting others in competition, when contemporary life is interdependent; to act as much as we can like efficient robots, when the initiative and spontaneity that are our human gifts are the keys to progress and success; and, as women, to take care of domestic tasks on the side while also doing everything men do in the public sphere at great cost to our health and happiness when what is needed by others is our wisdom. To manage these contradictions, today's heroines must protect their joy and aliveness above all. To do this, we need to live embodied lives, even though it is difficult to do so in the contemporary world, where many jobs require us to spend all day sitting, thinking, and engaging with our digital devices. Living in the body helps us come down to earth and recognize what is reasonable, real, and healthy for us, and what is not.

Sometimes our wild selves erupt in moments of spontaneous liberation. When the Dionysian rites came to town, women would jump up from what they were doing, throw down whatever they were holding, run out into the street, and follow Dionysus into the woods. Part of Dionysus's role is to shatter deadening structures, so this archetype is busy today disrupting all the old hierarchical, bureaucratic systems, which, as you may have noticed, are not working all that well. Women's lives are as complicated as they are because we have entered nontraditional fields while social structures and policies continue to reinforce the realities of an earlier time. The Dionysus archetype in us can help dislodge these structures by shattering our belief that there is nothing to be done about them; by getting us to talk about their cost to women, children, and families; and by helping us to spread the word about alternative systems that are working.

According to chaos theory in contemporary science, dissipative structures—that is, those that break up entrenched arrangements of all kinds—are essential in nature so that systems can self-create by developing new forms. Dionysus is the poster child for shaking things up to dispense with what is confining and oppressive in the old ways to make room for new realities. In between the shaking up and new life is a magical time of a creative void, out of which springs eros as a creative impulse that can then take material form (which is what Abraham means when he says that eros and Gaia are born out of chaos). When faced with plans falling apart, you can stop, breathe, and allow your intuition and imagination to let you know what wants to emerge in you and in your situation. This emergence is then like the sprouts of grass or even flowers growing between the cracks in the sidewalk.

Showing up as our emerging selves instead of hiding parts of ourselves to fit in or impress others shatters our false selves and helps to break down outmoded cultural norms and social structures that make us feel leaden, tired, and even depressed.

The answer to this is counterintuitive. The busier we are and the more chaotic things seem, the more playful we need to be. When we are at play, we tend be spontaneous, rather than obsessed with what others might think, or with regrets about the past or worries about the future. To connect with the world in this way requires that your authentic and deeper self be embodied in your physical presence, your actions, and everything you do, allowing you to act with spontaneity and to innovate in the moment. When you can show up in your fully embodied self, not only do you thrive at a new level, so do those you lead or influence.

In *The Four-Fold Way: Walking the Paths of the Warrior, Teacher, Healer, and Visionary*, Angeles Arrien notes that in many indigenous cultures, if you complained to a shaman of being depressed or unhappy, he or she would ask when you stopped "dancing," "singing," "being enchanted by stories," or being comforted by "silence."[1] If we all did more of such things, we would be happier and have more energy.

Dionysus is the god not only of joy but also of dancing and the other activities we do for enjoyment rather than to accomplish something. I remember first feeling the call of Dionysus at a time when my professional obligations were becoming more demanding, as they often do for women who simultaneously are, as I was, experiencing the sandwich-generation squeeze, balancing care for children and aging parents (or other familial or friendship responsibilities). When things seemed the most out of control and overwhelming, I would have a moment or more of panic.

Although my life overall was fulfilling and meaningful, I was focusing too hard on keeping going to hear myself think. Even when I was doing something that should have been relaxing and fun, my mind was always racing, stressing about work or personal responsibilities. Then I saw a *New Yorker* cartoon of a man in a bathing suit sitting at a big desk on the beach, working. I laughed, recognizing myself, for even when I took my kids to the park or another fun activity, some part of me was still back at my desk—figuring out what to say, what to do, how to analyze a pressing problem. I needed Dionysus badly. Although I did not believe I had the time, I suddenly felt the urge to sign up for a dance class. I didn't consciously connect this choice with Dionysus, although I did know that he was the god of dancing and embodied joy.

It has helped me to recognize that the root of the word panic is *Pan*, the half human, half animal god, so we can think of our response to those times of anxiety, with so much going on, as like a deer or other creature being startled in the wild and running around haphazardly. Dionysus was a later development from Pan, having morphed into a better model for humans of how to be in touch with their instincts, their body's wisdom, and their feeling of connection with nature, while also maintaining the ability to choose to be moral and thoughtful about what behaviors are harmful or helpful. Thus, from a Dionysian perspective, I could stop panicking and realize that the chaos in my life was awakening my wild woman and urging her to find a way to escape from the cage of my lists, plans, and those internal narratives that tell me I should always have things under control.

I'm sure I'm not alone in my awareness that my desire to keep my life under control was making me a tyrant, at least to my inner self, or how much I was longing to experience aliveness and vitality—and not just vicariously through watching a movie, viewing a sporting event, or reading a book. I've always resonated with Joseph Campbell's assertion that people do not yearn so much to know the meaning of life but for the experience of being fully alive. I know that I, for one, want both. I'm sharing the ideas in this section with you, just guessing that you might need them too, or if you do not, that you would appreciate getting validation for your fun-filled life.

Who among us does not sometimes feel imprisoned by all of our responsibilities and by the various ways we rein ourselves in so as to fit into current cultural ideas about what we should be to fulfill our various roles? Have you ever felt as if you just wanted to break out and connect back with the wilder part of you that you may have experienced as a child, or have never much been able to experience, but you know is there?

Dionysus is the lord of spontaneity, wildness, and exuberance, generally expressed in the relatively safe outlet of dancing. His rites called people to leave the constraints of civilization to dance wildly while satyrs would look on, clearly getting turned on by the experience. Satyrs were woodland creatures who combined animal (often horse or goat) and human qualities, similar to the god Pan. Dionysus's association with wildness and satyrs suggests that part of his role is to help us reconnect with our animal origins and wildness, as well as to physically spend time in, and protect, the wilderness. His rites were primarily for women, although some included men. A few were for men only, and some, for both men and women, celebrated the annual wine harvest. The joy he offers comes from a reconnection with your primal instincts and full aliveness. If you ever have felt that your life was too constrained, too responsible, too scripted, it is time to find the Dionysian impulse within.

Jungian analyst Clarissa Pinkola Estés, in *Women Who Run with the Wolves*, argues that modern women have been so overly domesticated that our fatigue comes from living lives too constrained and small. The instinctive "wild" woman is not wild in the sense of being out of control; rather, she is clear, decisive, and in touch with her body and its cycles, her emotions and what they are telling her, and her creativity and what needs to be created and what needs to be let go. She has strong boundaries, knowing when to say yes and when to say no. At the end of the day, her fatigue is from intensely exercising, creating, loving, and whatever else she clearly wants to do. Estés identifies the wild woman as the essence of female integrity, including being in touch with wildness as a "means to establish territory, to find one's pack, to be in one's body with certainty and pride regardless of the body's gifts and limitations, to speak and act in one's behalf, to be aware, alert, to draw on the innate feminine powers of intuition and sensing, to come into one's cycles, to find what one belongs to, to rise with dignity, to retain as much consciousness as possible."[2]

Female resilience comes from regaining connection with the inner wild, instinctual woman who has not yet been dampened down by her socialization. Estés once shared with me that she was homeless for a time, living in her car with her small children, yet she persisted in finishing her education, leading eventually to her becoming a respected analyst and a bestselling author. She credited her connection with the instinctual woman within her for her ability to succeed.

Some parents in ancient Greece would send daughters (who in our time would be in middle school, but then were soon to be married) to live in the forest and experience the freedom and power of the goddess Artemis—a virgin goddess, one-in-herself—who was free, independent, and also a huntress. There, the girls would learn to forage, hunt, and enjoy living in the wild together, bonding as girls. Because

Artemis also was a moon goddess, they learned about the relationship of the cycles of the moon to their monthly menstrual cycles, reinforcing Demeter's teachings about their connection with Mother Earth and her ways.

I've not been able to discover whether these were overlapping, competing, or the same puberty rituals as Demeter's; however, they would produce women who felt connected to one another and the earth, who were strengthened before they entered marriages that were quite role defined. They engaged in rituals to honor Persephone, preparing them to be sexually active and most often, married—or since Greek culture was not homophobic, they could choose a lesbian lifestyle, among other options. Just as I see the emergence of women's power in the Eleusinian tradition, Jean Shinoda Bolen (in *Artemis: The Indomitable Spirit in Everywoman*) finds it in the similar reemergence of the Artemis archetype in our time.

Even though women's roles in Greek life in ancient times were constrained, having them stay connected to their wild sides must have been a societal priority, or else they would not have been able to show up in droves at the call of Dionysus. Summoned into the woods by trumpets and drums, his acolytes would follow him into the forest and dance for days. Some (clearly apocryphal) stories even describe the maenads (Dionysian female followers) tearing wild animals apart and eating them raw, which would not actually have happened. (Would you want to do that? Ugh!)

Writing in the 1970s, Robert Bly, in his bestselling book *Iron John: A Book About Men*, observed a return of instinctual energy in women, which may have come from participation in the women's movement of the time. However, I don't see this wildness today, when most women are extremely well behaved (or maybe that is just me and the overwhelmed career women I hang out with). Bly was concerned that he did not perceive the equivalent in men and argued that they had gone too soft. His antidote was to connect them with their inner "Iron John," who is the instinctual wild man, analogous to the wild woman that more recently seems to have gone underground again.

The image of Iron John is powerful. He is glimpsed at the bottom of a pond, huge, muscular, covered with hair the color of rusty iron. This is a compelling image of the source of instinctual power, which hair often represents—which is why women in oppressive situations typically are forced to keep their hair covered.[3] Bly's hero rescues Iron John by calling together his comrades, who with buckets help empty the pond to free the wild man at its bottom. The symbolism here suggests that Iron John's power can become available to men—and, I'd add, women—if we can rid ourselves of repressed emotions, aspirations, and traumas, of which water frequently is a symbol. However, in Bly's telling, when the hero brings Iron John back to the king's castle, the king finds him so threatening that he keeps him in a cage, entrusting the key to the queen.

Although Bly rightly distinguishes this source of male power from the kind of macho-male culture that devalues and oppresses women, he says the hero metaphorically needs to steal the key from the queen. In Bly's men's retreats, they accomplish this by going off into the woods with other men to learn how to be free of an obsessive desire to please women (mother, girlfriend, etc.). Only then can a man offer himself as a worthy lover and equal to a woman (or another man).[4] A parallel to this is equally true for women, since we also need to work to free ourselves from the desire to please other women as well as men.

Gestalt psychology, with its emphasis on self-discovery through spontaneity and the power of living in the moment, and feeling what we feel and knowing what we know, can promote the Dionysian spirit in any of us. Dionysus also is the god of embodied living: being in touch with bodily sensations, our five senses, our capacity to sense pleasure and pain in the body, the wisdom that is gained from these, and the resulting power to take action.

Many of us know things about what would make us happier but don't act on that awareness. We may feel compassion when watching something touching or sad in the news, but we don't actually reach out to help anyone. We imagine doing great things, but instead sit on the sidelines and critique those who are doing them. We may engage in psychospiritual practices, gaining more and more familiarity with our inner lives, but never express the resulting richness in what we do. Demeter, Zeus, and Persephone cannot make us happy, prosperous, and fearless in their disembodied states. Dionysus brings human virtues down to earth, urging us to engage with life actively, and not someday, but now.

The archetype of Dionysus also calls us to the joy that comes from discovering what we love to do as play, or for pleasure, and also from feeling a sense of belonging with other people and with nature. I adore Mary Oliver's poem "Wild Geese," and find that many others do too. In it are the secrets to awakening the Dionysus archetype. First, the poem alerts us to the need to heed the wisdom of "the soft animal" of our bodies, then to notice the wild geese, calling us out of ourselves so that we can be filled with awe at the splendor of nature around us and recognize how fully we belong in this wild and beautiful world.[5] In that way, Dionysus is the cure for the alienation that so characterizes modern life.

The Dionysus Archetype

In ancient Greek art, Dionysus occasionally was depicted as an elderly man with a beard, and in this guise, he was connected to the kind of challenging wisdom associated with Socrates. More frequently, he is portrayed as young, very androgynous, even effeminate, often partly undressed and generally in relaxed, even languid poses. Although Dionysus has many sexual adventures with women and eventually marries happily, he also has been seen as the god of homosexuality. At least one myth tells of a same-sex relationship, and another tells of how, when Prometheus was molding the human race out of clay, Dionysus influenced him to give some a same-sex orientation. The presence of his exuberant spirit in gay culture, and in public events such as the annual San Francisco Lesbian, Gay, Bisexual, and Transgender Pride Celebration, has reinforced this association over time, as does the term *gay* itself. As a populist god, his rites always were open to everyone, and he carries some of this democratic spirit in his own nature, being radically inclusive and welcoming of diversity.

Moreover, the gifts of Dionysus are essential to meet the major challenges in our individual lives. Our heroines' journeys will, for sure, be less stressful if they are energized regularly by the joy Dionysus embodies, and his stories show us how we too can achieve the joy-filled life. As you have seen, each of the archetypal journeys builds on the prior one (if you skip lessons, you can go back and retrofit them), and Dionysus's journey is no exception. He likes to party, and as a god of women, it is interesting that he is credited with inventing the process through which grapes are made into wine, since wine, even today, often is the drink of choice for many women, just as beer and hard liquor are for men.

Dionysus also is linked with the vines themselves, and the need to prune them back.[6] Pruning also is a sign of his ease in letting go and moving on, an ability that is reflected in the story of his origins. But more than that, the image of the vines in Dionysian lore is similar to how Jesus used this image, with all people and groups potentially connected as branches to the same plant, which infuses them with spirit and the joy it brings.[7] The spiritual practice of Dionysian dancing was designed to help people experience collective joy through connecting with the spirit of joy in nature, one another, and themselves.

In the mythology of Minoan Crete, Dionysus is the son of Persephone, either by a virgin birth or impregnation by Zeus disguised as a serpent in her sleep, and most mystical readings of her story portray her as his mother, without necessarily identifying a father. Whatever is done to Dionysus, he survives or is reborn. In the most well-known Greek story, he is the son of Semele, a mortal royal princess of Thebes, and Zeus. Hera finds out that Semele is pregnant and convinces her to ask to see Zeus in his full glory, the sight of which instantly kills her. The Titans, who were aligned with Hera and threatened by Dionysus's revolutionary potential, destroy everything but Dionysus's heart, which Zeus places in his thigh, where it grows into a fetus that eventually is born through Zeus's leg.

In support of Hera, the Titans continue to do everything they can to kill Dionysus as he grows up. To protect him, he is raised in secret and often disguised as a girl. In some tales, they kill him repeatedly, but he revives. In others, they fail to find him. The story of Dionysus's dismembering is mentioned by Plato (in *Phaedo*) in a description of how Socrates linked Dionysian rites with what he saw as initiations in the philosophical path, which would have to mean dismembering ill-thought-out common ideas. More often, the dismembering and renewal relates to the capability of letting go and beginning anew or to a capacity for resurrection.

The problem with Zeus being viewed as Dionysus's father is simple. Zeus is credited as the father of almost all of the next generation of gods, Hera is blamed for everything that goes wrong, and almost every locality in the wider Greek world has myths in which they take credit for hiding Dionysus. With his capacity for undermining existing structures and questioning established ideas, Dionysus was a bit menacing to the status quo, which is a more logical reason why the Titans were threatened by him. While none of the major myths assume that Hades was Dionysus's father, nothing else fits with the actual Eleusinian narrative, with the clear similarity of their characters, or with Dionysus's ability from birth to move between the worlds, being at home in both the Underworld and the Upperworld.

Dionysus in the Eleusinian Rites

Dionysus is largely absent from the Demeter and Persephone story; yet his story is a focus of the Lesser Mysteries, which were held in February, where many came to prepare for the nine-day ritual where the deeper secrets were shared. His birth is celebrated in the final, most sacred and secret part of these Greater Mysteries. In this way, his archetype serves as the fruit of the vine, or the icing on the cake that completes the lessons of the other archetypes. For even if you have achieved great things and have been wonderfully caring and altruistic, as well as romantic, you can lack a capacity to experience joy and the ability to be uncensored and spontaneous, present in the moment. That is why we all would be wise to awaken Dionysus's gifts.

Dionysus was a prominent figure in the Eleusinian procession, and while reading this, it can be fun to envision an entire horde of pilgrims walking together along a path, demonstrating joyous revelry. On the fifth day of the festival, initiates made their way from Athens to Eleusis, accompanied by a statue of the boy-god Iacchus (Dionysus). We can imagine that this might have been on a ritual cart or open chariot, pulled by an ox or horse and accompanied by dignitaries. And many in the procession would carry rods made of woven branches and a myrtle wreath as a celebration of him, and they would cry out joyfully, "Iacchus, O Iacchus," which scholars agree was another term for Dionysus, since different localities had different names for him. Initiates would dance in Dionysian ecstasy, whirling, accompanied by every kind of musical instrument (which most likely would have been provided by eager musical participants), and yell out cries of celebration, sometimes adding vulgar jokes in honor of Iambe, who was the servant who broke through Demeter's grief and got her laughing.[8]

When individuals would walk one by one over a narrow bridge, hooded figures would single out well-known, possibly pretentious people to target with satirical insults; in doing so, they honored Iambe, but also Dionysus, who was famous for having no use for pomposity. This tradition was a source of great merriment and social leveling. In this way, Dionysus served the same function as the fool in the ancient medieval court who used humor to keep the royalty from getting too full of themselves. And because the narrowing of the road would slow things down, it might have had the feel of lines at Disneyland, except those in them had to entertain one another, since they had no virtual entertainment, just themselves.

Between the dancing, and the excitement of the ritual events, and fasting on the sixth day after their long walk and in preparation for the deeper, secret parts of the rites, people would be in an altered state, even without the use of drugs, experiencing a kind of Dionysian ecstasy. However, some scholars believe that a hallucinogen was employed to create the life-changing impact that was reported, a possibility that is a major continuing debate. This final evening had elements that were chaotic, with initiates milling about, shouting, and jostling one another, which in such a large crowd and in the dark also was disorienting. The initiates' confused wandering was done in honor of Demeter's wandering, but it also celebrated the power of chaos as the generative void out of which new life begins.

The initiates finally would arrive at the entrance to the Telesterion, where the secret rites would be held. The password required for entry was some version of the following: "I fasted; I drank the *kykeon*; I took from the *kiste* [a metal cylindrical container]; having done my task, I was placed in the basket, and from the basket into the *kiste*."⁹ The kykeon is the barley water that Demeter drank instead of wine and that might, or perhaps not, have contained a hallucinogen. (I tend to think not, because all the ones that have been suggested are dangerous, and initiates dying from an overdose likely would have undercut the attraction of the rites.)

The rest of the password suggests that initiates already had honored Dionysus in some ritual way that involved these elements. The latter part is thought to be a reference to Dionysus's resilience when the Titans were trying to kill him. As an infant, he was hidden in a container, taken out in secret, and then placed in a winnowing basket, which was used as a cradle. He may have been placed back in the kiste when it was time to hide him in a new location. Since winnowing baskets are used to harvest grain or grapes, this would connect him with the Eleusinian theme of the agricultural cycle and particularly that of the grape, which was sacred to him. Moreover, during the most secret parts of the rite, where the magic of the sacred union of the god and goddess would have occurred, a bright light suddenly blazed up from the fire, shattering the darkness. According to Hippolytus, at this moment a hierophant (priest) would shout, "Brimo has given birth to Brimos!" Brimos was yet another of Dionysus's names, a word that also means "thunder," since Dionysus was known for his loud, thunderous voice. Likely shocked by the sudden and startling light, initiates surely would exclaim, creating a thunderous moment.

Some scholars have made a case that the birth being celebrated was Persephone's, not Dionysus's, but my sense is that it may not have mattered which it was because the essential element is the reinforcement of a sacred birth. Both Persephone and Dionysus are important to the rite; that much is clear. Scholars concur that the secret parts of the

Eleusinian rite also included a celebration of the birth of some divine child. Ovid (in *Metamorphoses*) praises Dionysus as the *puer aeternus* (or "eternal child") of the Eleusinian Mysteries, viewing him as a god of vegetation and resurrection.[10]

In numerous spiritual traditions, such stories, of an infant seemingly small and weak—the Christ child, for instance—affirm the positive potential of the future, and in you and in me, as well. For the initiates, it likely was a celebration of the birth of the divine in them and their now being adopted by Demeter and Persephone. That also may be how Dionysus (the god) can give birth to Dionysus (in the initiates), which would explain why it is Brimo, singular, who gives birth to Brimos, plural. Since he is the god of joy, this birth would suggest the awakening of happiness, expressed in loud rejoicing, within them.

The lessons that follow take us deeper into what it means to awaken or strengthen the Dionysian archetype within us. As you may have noticed, each section of this book has been both additive and integrative, so that by the end of part 2, Zeus now balances his qualities with those of Demeter; and by the end of part 3, Persephone has built on the qualities she derives from her parents, Zeus and Demeter, by adding her own.

Part 4 shares Dionysus's archetypal lessons, and also shows how all four archetypes come together.

+ Lesson 1 addresses specific myths about Dionysus that help us understand how the Mysteries result in happiness, prosperity, and freedom from fear.

+ Lesson 2 utilizes a contemporary film to explore what is needed for individuals to upgrade fun into genuinely fulfilling joyous experience.

+ Lesson 3 considers the power of collective dance and celebration to move us beyond alienation to radical belonging—with others and the earth.

+ Lesson 4 explores how we can be the authors, directors, and actors of our own dramas.

The Dionysus Archetype
Mythological Dionysus: the god of embodied joy, dancing, wine, and ecstasy
Primary Heroine Lesson: being present in the now, savoring life in the body
Narrative Progression: from madness to communal joy bringer
Gifts: joy, freedom, moderation, spontaneity, presence, and healing
Historical Gender Association: masculine androgynous, friend to women
Decision-Making Mode: kinesthetic bodily response
Inner Capacity Developed: the wisdom of the body
Counterproductive Forms: addiction, seven deadly sins

Dionysus and the Eleusinian Promise
Primary Happiness Practice: living in the present and gratefully savoring life's pleasures
Contribution to Prosperity: fully savoring life and its treasures while feeling radically alive
Contribution to Freedom: living in the moment, free of regrets or fears about the future

APPLICATION QUESTIONS:

Based on what you know so far about Dionysus,

Do you have too much, too little, or not enough of this archetype?

If it is present in your life, was it a family legacy (i.e., you were taught to be like that and are), a vocational ally, or something deeply and authentically you?

Do you like people who reflect Dionysus's stance in the world? Why or why not?

Dionysus Lesson One:

Realizing the Eleusinian Promise

INITIATES AT ELEUSIS FELT so much happier after the ceremony, not only because of what was taught but also because they had an experience that completely altered their worldview and how they felt in their own skins. Dionysus challenges us to gain radical aliveness through experience, not just mental knowledge, and with others, not just alone or in pairs. This lesson focuses on Dionysus and myths about him that contribute to the realization of the Eleusinian promise: happiness, prosperity, and freedom from fear. These stories also illustrate two very important life lessons that help us fulfill this promise: avoiding excess and discovering what god is calling us when we exhibit symptoms such as lethargy or depression.

Heroines take responsibility for being the best they can be, but they are not focused only on themselves. What differentiates them from ordinary women is their concern for others and the larger social good. That is why the Eleusinian promise is addressed here at both the individual and collective levels. Let's begin with Dionysian teachings on the promise of prosperity, since so many people today confuse money, which is just the tool of our barter system, with prosperity itself.

The well-known tale of King Midas is a Dionysian teaching story about the ills of immoderate behavior as applied to prosperity. According to some sources, Dionysus gets Silenus, his mentor, drunk so that Silenus will share his wisdom with him, since Silenus was known for his sagacity. Silenus then wanders about and passes out in Midas's rose garden, and Midas takes him in and listens to his amazing stories. Feeling responsible for Silenus's plight and grateful to Midas for sheltering him, Dionysus offers to grant Midas one wish. However, when Midas asks that everything he touch turn to gold, Dionysus warns the king that this is not a good idea, but Midas insists.

At first, Midas is thrilled, as he turns this thing and that into gold. But then, when he tries to eat, his food turns to gold, and when his daughter comes to give him a kiss, she turns into a golden statue. Of course, Midas begs Dionysus to remove this curse. Dionysus assures him that he will be freed of his wish and its consequences by bathing in the river Pactolus, which then, because its sands turn to gold, becomes a source of gold accessible to all.

This story actually warns against two types of excess. To get drunk is to not know when enough is enough. Had Silenus not been taken in, he would have been vulnerable to harm by others—plus, of course, drinking to excess is unhealthy. Many things may be fine in small doses, but in larger doses they end up making people miserable and potentially ruining their emotional or physical health. Gold (i.e., money) is a means of exchange that can help you get what you want. It can buy experiences that fulfill you, like education or travel, and beautiful and serviceable surroundings, and allow you to give yourself and others an abundant and prosperous life. However, you cannot eat it, have sex with it, or have any real experience with it. You can love it, but it does not love you back.

By itself, money will not make you happy, as any number of prominent but miserable people reveal to us daily. Turning his daughter into cold, hard metal is the horror that wakes Midas up. This cautionary tale is not just a warning against avarice; it is an admonition not to monetize everything and think of ourselves, our children, and our employees as commodities or resources. Of course, having enough money to fulfill our destinies and enjoy our lives is important, but many people have plenty of money and feel it is never enough. Real prosperity is about what it takes to generate plenty and then have a genuinely rich and rewarding *experience* of abundance.

So how is prosperity attained at the societal level? Each archetype has a role: The Zeus part of us fosters the work instinct, the desire to make a contribution, and the motivation to gain mastery so as to give our best selves to the whole. The Demeter part of us furthers this by motivating us to care for one another and ourselves, to give a leg up to those falling behind and a welcoming hug to those feeling excluded. Demeter also provides lessons from agriculture to help us partner with others and the earth to increase our collective yield, and she encourages us to willingly contribute (say, through taxes or charity) to provide for those in need. Persephone provides collective meaning to actions that bind us together and encourages the formation of the intimate romantic and friendship bonds that stabilize and weave the fabric of society, while also promoting creativity and innovation, thus furthering progress. To this, Dionysus adds all the ways we celebrate as communities, so that we not only have abundance but also as a society can savor what we have and be grateful for it.

On a personal level, Zeus helps you find and hone your unique gifts, accomplish your vocational and financial goals, and put in the work required to attain them. Demeter augments this by motivating you to care for yourself, those around you, and your possessions to avoid costly breakdowns. Her altruism enables you to experience abundance through generosity, giving away what you do not need. As the Lord's Prayer in various translations reminds us, we will be forgiven our "trespasses" and "debts" to the degree that we forgive those of others. Generosity of spirit attracts generosity back. Even those in direst poverty who help one another fare better than those who do not. This was made evident by the Fourth Ward in New Orleans, where people with little education and money led decent lives because they helped one another—that is, until Hurricane Katrina destroyed their neighborhood and the surviving residents were evacuated to different parts of the nation, which left them without their support networks.

Persephone puts you in touch with your intuition, so that you can generate the new and innovative ideas so critical to making money. She teaches you how to be guided by eros in how you then spend that money and time—husbanding your resources so that what you have and do fulfills you. And Dionysus protects you from the scarcity mind-set that can keep you from enjoying all the riches available to you—everything from your financial resources to a brilliant sunset.

A pernicious form of talent scarcity has been the way traditional sex roles have cut people off from developing their full human capacities, just as unequal educational and other opportunities have disadvantaged many in poverty, preventing them from becoming all they could be. Dionysus illustrates the advantage of androgyny, since he did not have to limit his behavior to fit the standards for manliness of his time, and thus increased his inner treasure trove of available capabilities. From this, he was able to create his own rite, be a star at Eleusis, and attract a following of women because he understood them. Having been raised in hiding, dressed as a girl, he sported an androgynous look, even when as an adult he dressed as a man. Consequently, he had no need to repress his feminine side; because he knew what it was like to be treated like a girl, it was easy for him to be a good friend to women. And having been separated from his family, living in hiding and always moving around, he identified with others who had had such a rough upbringing, which is why his rites, like the overlapping Eleusinian ceremonies, might have met the needs not only of women but also of men and women from the lower economic caste.

Dionysus's honoring of women also is revealed in his strong and faithful marriage to his wife, Ariadne.[1] Because Ariadne is a mortal, she eventually dies. After her death, Dionysus travels to the Underworld to retrieve her, then renders her eternal in the form of the constellation Corona. Marriage partners in Greek myths often serve as metaphors to reveal something about the god in question. If you remember the story of the Minotaur (in Zeus lesson 3), Ariadne is the Minotaur's sister. It is she who understood how to traverse the labyrinth—the same woman who first married and then was abandoned by Theseus, the slayer of the Minotaur. Not surprisingly, being abandoned by him on an island left her disillusioned with hero types who use bullish power for conquest. Dionysus found her on the island and immediately fell in love with her. Although by this time the Minotaur is dead, he symbolizes the capacity to tap into your bullish strength either to gain power, first as Theseus does, and then to develop the stamina to express your wild and risk-taking side, as Dionysus does.

Dionysus's capacity for resurrection or regeneration thus helps him realize the Eleusinian promise of freedom from fear. Since he was killed and resurrected multiple times in the womb and then as an infant, it was credible to believe that he allowed himself to be torn apart (at least symbolically) in every rite as his followers, the maenads, in the throes of ecstasy, grabbed parts of him, like modern fans grabbing for rock stars' clothing. By engaging in such uninhibited dancing, his followers released frustrations, angers, and lost dreams through a cathartic experience that enabled them to die to their current unhappiness and be reborn in bliss.

In the ancient rites, Dionysus is present as an energy infusing joy into those invited into the dance, while also helping the dancers learn the secret of dying every moment by letting go of the past or of worries about the future in order to let in the pleasure of the dance itself.[2] Dionysus wants us to stop addictively holding on to possessions, to any sense of what we have to be, and even to our desires and dreams. All these can be held lightly, so that we can let go of them if attachment is causing us to suffer.

Again, the other archetypes provide a foundation for this. From Zeus, you gain the confidence to declare independence from what others think and want from you and to take risks to realize your dreams. From Demeter, you learn to accept the cycles of life, to protect and nurture whom and what you love, and to grieve necessary losses and move on, having faith in the natural and spiritual processes of life. From Persephone, you learn to focus on the positive, make informed choices of what you want, and use your creativity to make the best of life's opportunities and challenges. Dionysus keeps your attention on savoring life and appreciating people for what they are, rather than feeling the need to change them—which does not mean that you cannot still remember what you are striving for in your life. You just are not swept off course by the negativity of others or your annoyance at them. I once heard a story about a vendor who sold newspapers on the sidewalk. Every morning, the same man came by, bought a paper, and treated the vendor with disdain and verbal abuse. The Dionysian vendor would just give him a big smile and say, "Have a wonderful day." When asked how he could respond that way to someone who treated him so badly, he explained that he did not let anyone take away his freedom to be happy and kind.

Dionysus actually went further than this, since he believed that the antidote to unhappiness and fear was ecstasy, checked, however, by common sense. The two foundational premises of ancient Greek morality and preventive health were Socrates's famous injunction, "Know thyself," and Aristotle's doctrine of "the middle way," which advocated all things in moderation, nothing in excess. Dionysus often is mistaken for Bacchus, the Roman god of wine, known for drunken orgies. Although Dionysus was the god of wine, Jungian analyst Robert Johnson, in his book on Dionysus (*Ecstasy: Understanding the Psychology of Joy*), argues that Dionysian initiates would have had enough wine to overcome their resistance to letting go into the dance, but there was no sign of excess in the historical records.[3] Dionysus did not need external sources of energy; he was in touch with the bull within, so much so that he could share that energy with his followers. He taught how to be inebriated by life, love, and the spirit flowing through us. In this, Dionysus was carrying on an ancient female shamanic tradition of trance dancing, which allowed spirit to flow through the dancers and teach them what they individually or collectively needed to know to thrive. His path, then, supported by the other gods and goddesses, furthers health and provides healing when illness occurs.

Dionysus was driven mad early in life by the Titans' constant attempts to kill him. He traveled widely in search of a cure, and eventually regained his sanity, as well as a variety of gifts, including comfort with apparent chaos and the ability to invoke ecstasy. Since he was restored to health by a Mother Earth goddess similar to Demeter, we can assume that his cure involved acquiring a mind-set of connectedness to the universe, the earth, and other people, coupled with an awareness of personal responsibility to make a difference, because after that point he seemed clear about his mission.

The combination of his Zeus and Persephone attributes would help him discover that his calling was to bring the ecstatic into the world, and with it, the radical health and freedom that so threatened the other gods that they wanted to kill him. Based on his experience, Dionysus came to see madness as a sign that someone was ecstasy-deprived. He had no use for puritanical types who tried to stop his rites or those who wished him harm. His myths report that he would inflict insanity on either the offending joy killer or a nearby woman. If the latter, she likely would tear the uptight person apart. Remember here that these are myths and meant to be read metaphorically. Puritan types generally are rigid and judgmental and want to keep others from having fun. It could be a service to them to dismantle the inner chains that keep them from joining the dance.

The spell Dionysus casts on his detractors that strikes them with insanity is not a punishment; rather, it is a way to cause needed radical change. The Greeks identified numerous forms of madness, many of them positive; these included *hubris* (which was so dangerous that it threatened social and societal breakdown), *prophetic* (as with the Delphic Oracles), *telestic* (initiatory or ritual), *poetic* (as in possession by the muse), and *erotic* (as in romantic love and mythical union with the divine). We can imagine that Dionysus's initial insanity emerged as a compensation for his despair and manifested as hubris—possibly leading him to become inflated with his powers, given his rather arrogant belief that he should take it on himself to save people who did not like his Mysteries. Once he is healed and comes down to earth, which he does through participating in Demeter-like rites, he gains access to multiple forms of positive madness: the telestic in ritual dancing, the poetic in his inspiring Greek drama, and the erotic in how those who participate in his Mysteries feel that they are unified with the spirit infusing the universe, as represented in his form or those of other gods.

The Greeks identified a special form of Dionysian madness as that which disrupts hardened ideas that make people unhappy and rigid and allows for a psychological reconfiguration that enables them to attain greater happiness. Thus, Dionysus both causes madness and then cures it, but only when the person's rigidity, addiction, or compulsion is understood and addressed.[4] This notion is similar to Hillman's famous proclamation that "all our pathologies are calls from the gods."[5] When we learn what an archetype wants from us and we do it, we are likely to be healed. That is why Hillman, in *Re-Visioning Psychology*, tells us that when we have an image in a dream, rather than analyzing its meaning, we should ask it what it wants of us. Otherwise, it is likely to keep trying to get our attention through unpleasant symptoms.

Dionysus typically offers ecstasy or, to those who refuse it, madness, as a means to break down rigidity—so that they can open up to the bliss of being. Those who offend him die to what they have been before. Their defenses are torn away so that they can experience the divine ecstasy that is Dionysus's ultimate gift. In one story, after pirates capture Dionysus, grapevines grow from the water and take over the ship. Out of terror, the pirates jump overboard. Dionysus then changes them to dolphins, creatures known for being both intelligent and joyful, with swimming as their form of dancing. If we interpret this story metaphorically, we might imagine the pirates gaining the human equivalent of such dolphin qualities (in addition to the ability to thrive in the sea so that they do not drown).

Even today, understanding the message of a symptom can restore hope to those suffering from persistent illness, with the result that they experience an upwelling of quiet joy. Alternative medicine scholar Bonnie Horrigan, writing about the healing power of discovering meaning in experience, describes the previously mentioned practice of dream incubation in Greek life, a practice that was associated with Asklepius, the god of medicine and healing. She explains how dreams offer a commentary on the meaning of the symptom and the path toward healing:

Dreaming . . . was the one consistent feature in medicine all over Europe for more than a millennium. From approximately 1300 BCE to 500 CE, in Asklepian temples, physicians guided patients with seemingly intractable health problems through intensive healing retreats. A wide range of therapies was invoked but the capstone of these efforts was incubation, a period during which a patient fasted, prayed, and slept in an enclosed chamber called an *abaton*, which literally means the "inaccessible place." Through dreaming in the abaton, they would come in contact with forces otherwise inaccessible to their ordinary awareness.[6]

Dream analysis as an aid to psychological healing is standard in depth-psychology practice, which views dreams as letters from the unconscious to the conscious mind, revealing what we are not seeing, whatever that is. Social matrix dreamsharing is now common at conferences to capture what the unconscious mind is revealing. Horrigan notes that at least one contemporary center, founded by Jungian analyst Robert Bosnak, is reviving the tradition of incubation in order to complement medical interventions through the healing power of dreams.[7]

However, healing insights also can come from the conscious mind, for example, in response to the driven quality and resulting anger that infuses society today. When we are inexplicably angry or full of what seems to us like righteous outrage, we can enlist Dionysian insight to identify the healing potential in the symptom, and thus recognize that our discomfort is about the lack of what we genuinely yearn for and focus on finding it. I know that the harder I'm working, the more I resent anyone who seems to be taking off for the beach when important tasks remain to be done and seems oblivious to the cost to me and others of being left with their work as well as my own.

Dionysus invites us to turn and return to a path of radical freedom, of being fully and spontaneously ourselves—including living our desires and facing our fate—without stressing about the consequences of doing so. A balance with the other archetypes keeps us from going too far, which we are most likely to do when our lives have been joy deprived. For example, a pattern of self-denial can be the cause of someone suddenly rebelling and doing whatever they want without thought to the financial implications, their personal relationships, or their responsibilities. The women who ran out to dance with Dionysus were living very role-defined lives. He gave them the chance to feel totally free for a while. It was a release, like a very uninhibited vacation, not an excuse for lifelong self-indulgence.

Our personal liberties today are expansive. Freedom of speech is a protected right in the United States and many European countries. We cannot be imprisoned or fined for looking exuberantly happy or greeting the day with optimism and joy. However, our desire to be taken seriously often keeps us from showing how mad, sad, or happy we really are. Most of us try to be and look like good, responsible people, and the good-girl, always-be-nice-and-don't-make-waves mandate, even when self-imposed, is very hard to escape.

Without a balance with Dionysus, the former three archetypes also can be complicit in keeping us in this trap. Zeus energy creates laws and rules that tell us the right and wrong ways to do things; Demeter values enjoin us to help others before all else; Persephone's focus on authenticity can keep us obsessively soul-searching, not living. On top of that, we are subject to the ongoing pressure to abide by the rules and values of our society, our families, the crowd we hang with, the organizations we are part of (workplaces, houses of worship, the local hangout). However, all these pressures to be good boys and girls can begin to feel as if we are in prison. When that happens, we need a very big dose of Dionysus to stretch the boundaries enough to start doing what brings us delight.

The powers of Zeus and Demeter can crowd out Persephone as well as Dionysus. When we were children, our parents needed to socialize us to behave appropriately for the situation, and that required reining in less socially acceptable, or even dangerous, forms of authentic expression. Perhaps this is why children, especially in this age of helicopter parents, so love the movie *Frozen*, which is the highest-grossing animated film of all time. In particular, they adore the scene where Elsa, the Snow Queen, who has been bottling up her emotions in order to avoid freezing the things around her, sings "Let It Go" as she creates an entirely frozen environment. The lyrics are touching in how they convey the cost of repressing who you are and what you are meant to do, and the resolution happens when she declares that the perfect girl she tried to be is now gone—clearly a statement that touches the heart of a modern issue for adults as well as children.

Just as the Zeus archetype helps us access the energy from the inner bull, Dionysus helps us tap into the gifts we could learn to value if we just "let it go"—meaning all the powers we have that we are stifling in order to fit an image—at least long enough to know what our powers really are. Then we can find ways to express them that do not cause harm to us or anyone else. The plotline of *Frozen* reveals that balancing authenticity with appropriateness is aided when someone else loves us as we are. It is Elsa's younger sister's unconditional love that transforms her, allowing her to control the expression of her gifts so that she can apply them in socially acceptable, nondestructive ways. Sometimes we just have to find that younger sister inside ourselves, so that she always can be with us.

Being truly happy requires that you use the gift of Dionysus to express your exuberance when it has been repressed for safety and virtue's sakes. For example, say that you are trapped in a catch-22 situation where you feel that you just cannot win because anything you do will break someone's rule or result in someone feeling hurt or upset. Dionysus may show up as desperation and frustration so intense that you simply throw your hands in the air and give up trying to do everything perfectly and keep everyone happy. Generally, this is when clarity starts bubbling up from the body about what you really need to do at that moment to feel freer and less disempowered.

In some mythic art, we see that the Dionysian dancers were equipped with whips. We might interpret those whips not just in the traditional way, for holding off the lustful satyrs, but also as the ability to exert control so that we can express our wildness without becoming less than fully human or doing anything injurious to others or ourselves. We also might see them as the strategies we can use to create boundaries in case some of our gifts give others the wrong idea about what we might be up for. These boundaries even can be employed to curb the archetypes themselves.

The shadow sides of all the archetypes reveal a desire to take over the lives of mortals, expecting them to dedicate their lives to expressing the archetype's goals. This is one of the reasons that women are so exhausted today. Demeter wants us to be the perfect mother and to care for everyone around us who needs help. Zeus wants us to be wildly successful and wealthy. Persephone wants us to be soulful and deep, connected with the mysteries of the universe. And Dionysus expects us to be experiencing and fostering ecstasy all the time. Our inner whips are symbols of our need to develop ways to defend our personal identity against the archetypal demands on us, so that they can become our allies, not our masters. If we are unaware of archetypes in us, they can possess us in their more negative forms; recognizing them reinforces free will.

At their best, the four great archetypal Eleusinian stories can contribute to defending our freedom: Zeus, by helping us to know who we are and what we want, and by giving us the audacity to go for it; Demeter, by allowing us to care for ourselves as well as others; Persephone, through teaching us to exercise free will to choose our destinies; and Dionysus, by tempting us away from busy lives to take time to frolic.

The lesson that follows offers some idea of what Dionysus might be like if he were an individual today who is going through the motions of having a great time, while feeling blue and seeking to cheer up—in other words, someone who is living an inauthentic life. He has a Dionysian lifestyle, but needs to awaken the Eleusinian archetypes more fully to realize their combined promise.

APPLICATION EXERCISE:
Realizing the Eleusinian Promise

Think back on the times in your life when you have felt any one or all of the elements of the Eleusinian promise of happiness, prosperity, and freedom. Notice the role "the middle way" can play in helping you avoid both self-denial and greed. Where have you had a comforting sense that "this is enough; I do not always need to be driven to have more and better"? Then harvest some insights from what you have identified to help you savor the happiness, prosperity, and freedom you have right now, and imagine how you can continue to maximize these in your life going forward, while avoiding excess and deprivation in all their forms.

Dionysus Lesson Two:

Celebrating Life's Great Beauty

M ANY OF US ARE SO PREOCCUPIED with the business of our lives that we lose the capacity to let down our hair and enjoy leisure. On the other end, there are those who fritter their lives away playing all the time and find that they no longer get any pleasure from what they are doing. Yes, it is entirely possible to spend your life partying and dancing and not be happy at all. And, of course, if these activities lead to dissipation, they can produce a dangerous downward spiral. Crucially, Dionysian rites were infused with spiritual meaning; they were not simply recreational. This lesson considers how we can restore aliveness if we are too into status, being important, or trying to be perfect at everything to relax and enjoy what we are doing, or when we are out of balance so that our Zeus and Dionysus masculine modes are crowding out our Demeter and Persephone ones.

Some years ago, when I had the midlife blahs, I was drawn to films like *Zorba the Greek*, where I fell in love with the image of Zorba dancing on the beach, and *Shirley Valentine*, because I identified with the title character, who escapes a very drab and boring life to reclaim her aliveness on the incredibly beautiful Greek seacoast. My life was not drab, just so busy that I was feeling trapped, my body closed down, and my mind yearning for beauty. So I redecorated my home so that it was more attractive, and then began exploring various dance forms over time, including authentic movement, SynergyDance, Nia, and the 5Rhythms, that free up the body, link energy systems, unblock frozen areas in the body, promote ease with the various rhythms of life (from gentle and flowing to sharp and fierce to chaotic and wild to centered and still), and provide catharsis of difficult emotions for a natural emergence of more positive ones. It was a self-structured life course on learning to live in my body.

More recently, when I was feeling a tension between focusing on living fully and the long hours at the computer needed to write this book, I determined to break more often to go to dance classes, where I could not only move but also feel synergy with the group. However, my most powerful Dionysian epiphany occurred when I saw the Italian movie *The Great Beauty,* which received the 2014 Oscar for Best Foreign Language Film at the 86th Academy Awards. As I watched, I found myself thinking, about the film's main character, "If Dionysus were here today in mortal form, Jep would be he," except he is going through the motions of a joyful life but is not actually happy. And while the details of my life are very different than his, the resolution of his story, and the understanding he comes to that restores his excitement about life, speak to the part of me that can turn an inspiring calling into a must-do deadening routine. Perhaps this story will speak to you too.

The protagonist of the movie, Jep Gambardella, is a man who seems to have it all. He clearly is affluent and hangs out with high society in Rome. He built his reputation with an award-winning first novel, he is a respected art and society critic, and he spends his evenings partying. There appears to be no shortage of beautiful and interesting women who want to sleep with him. And yet, he is depressed. Like so many people we hear about or know personally, he possesses the external qualities that comprise a "successful" life, without the inner wherewithal to enjoy them.

The Great Beauty evokes the Dionysian spirit through its gorgeously artistic cinematography and its sound track, as well as through a succession of seemingly random images that feel chaotic but add to the emotional experience of the film. Viewers are pulled in two directions: empathizing with Jep's depression while also thrilling at the majesty and artistry of the setting and the mystical quality of the music.

The film begins with scenes of women singing from the balcony of a cathedral as the city awakes, followed by an assortment of what seem to be random events, including a tourist photographing the scenery who suddenly drops dead. We never see or hear anything more about him. Images of innocent children playing recur throughout the movie but have no role in its plot. All this evokes Dionysus as the god of chaos and randomness. Many additional scenes reinforce the themes of death, innocence and its loss, and the fleetingness of time, inviting viewers to feel their way through the movie without much of a clear and linear plotline.

The hauntingly beautiful music of much of the sound track creates a mystical, emotional field, speaking to the soul of the viewer. Some of this music emanates from the Catholic cathedral that is visually highlighted throughout the film. The music is so sublime that it made me think about a friend of mine whose beliefs are very different from modern, official Catholic theology, but who says she still is a Catholic. When I asked her to clarify how these two statements hung together, she replied that her faith is based on the deeper message of liturgical music and the beauty of much of its imagery. That music can speak to any of us, whatever our belief systems. Equally powerful in the film are scenes of Rome that are so beautiful that they create an inner tension in the viewer between empathy with the main character's angst and the transfixing loveliness of the setting and music.

In the movie, the musical mysticism is juxtaposed with modern ecstatic dancing, where one is portrayed as spiritual and the other as hedonistic exuberance. Jep is a quintessential Dionysian character who aspires not just to be part of the "high life" of Rome, which is characterized by all-night wild dancing and partying out in the open or on rooftops. Rather, he wants to "own it" by being able to determine whether the parties were successes or failures, which he can do as the art critic and social reporter for a local news outlet. It is clear that he has succeeded in reaching this goal, since people defer to him. This makes him the lord of the dance, the closest a mortal might get to being Dionysus, with a bit of Zeus desire and capacity to rule thrown in.

Yet at this point in his life, dancing and feeling important are not making him ecstatic. What narrative intelligence lessons can we take from his example? If you were he, you could use your knowledge of Eleusinian archetypes to figure out which ones are underrepresented in your psyche and which are expressed in their more undeveloped, shallow, or shadow forms. You might realize that this dose of Zeus dampens Dionysian joy, and that Jep is cut off from his Demeter caring for others and his Persephone connection with his deeper self.

Like so many people today who feel blue, Jep's first response is to focus on what is wrong in the world and with other people, and to express outrage about it. As currently is fashionable, he is good at leveling Zeus-like criticism and Dionysus-like anger at others. He can cut to the quick anyone who seems inflated or pompous, which we see first when he conducts a ruthless interview with an artist, and later in conversation with a friend. He has no difficulty recognizing the superficiality of the stilted and often phony life of the rich of Rome.

Jep is equally ruthless with himself, as he faces head-on how he is going through the forms of celebration without feeling much of anything. However, he is wise to his own faults and looks for help. With friends, he refers to himself, as well as them, as cynical, despairing, and without much merit. He tries to talk with a cardinal, who, according to rumor, may be the next pope, about his spiritual crisis. However, the cardinal, whom he meets at a party and afterward at dinner, is interested only in conversing about gourmet recipes and avoids spiritual discussions, while clinging pompously to his status in the Church.

Two characters compulsively take selfies, one of whom is a woman Jep walks out on after perfunctory sex; it appears that they take these pictures because of their need to see themselves in order to have any sense of who they really are. Later, when Jep attends an exhibition of pictures of a young man, one for every day of his life (first taken by his father and then by the young man himself), it becomes clear that by taking such photos, people are trying to find and know themselves. These scenes reveal Jep's need to gain deeper self-awareness.

Jep demonstrates masculine strengths, but his lack of feminine ones also is an issue for his entire society. The sad state of the feminine is illustrated repeatedly in the movie—by his friend's bitchy girlfriend, by a saintly woman who is so self-sacrificing that initially she seems pathetic, and by a beautiful stripper who is wasting away from an unknown disease. Jep's wise boss is the most confident female character in the film, but she is a dwarf, and tells him that this is what people notice when they see her. Yet often it is the women—glimpsed by or in relation to him—who help him recognize his narcissism and wake up his dormant sides.

As Jep sinks more and more into awareness of the gulf between who he is inside and his chosen roles, he observes an Arab woman with beautiful, sad eyes, which are all you can see of her. The implication is that he feels like he is wearing a burka, only his "burka" is a persona that hides who he is inside. Then, at a cathedral, he sees a young girl, hiding in a hole, who tells him that he is "nobody," undercutting his identification with being a big shot in high society.

Jep has entered a new stage of life that will require more of him—to experience more empathy and care for others and shed his defenses to show up more authentically. At some point, most of us experience a call to a deeper level of living through some existential crisis in which we come face-to-face with the reality of our mortality, which also pressures us to reveal our true natures more fully than we have. As we enter this process, we begin to experience the shame of all the ways we have been inauthentic, grief about times we did not grab the golden ring offered to us, and remorse for times when we have caused harm. For those who put this confrontation off until they are at an age when others may be retiring (as is the case for Jep, who is turning sixty-five), the call can be intense and urgent.

Sometimes dormant archetypes wake up just when they are needed, so that the changes required occur organically. (They also can be the result of choice, if you figure out what is missing and "fake it until you make it" while consciously letting go of any resistance to living previously ignored archetypal energies.) Out of Jep's pain comes increasing empathy, allowing him to open his heart as he acts out of his Demeter caring impulses. The people who are most important to him now are not the successful, glamorous ones but those who, for one reason or another, are confronting difficulty. For example, he befriends Ramona, who works as a stripper in her father's nightclub to pay for treatments that may keep her mysterious illness at bay and who scandalizes his friends by her revealing outfits, but whom we gradually discover is one of the deeper and wiser characters in the movie.

During this period, the son of an acquaintance of Jep's commits suicide, and before taking Ramona to his funeral, Jep cynically explains to her the rules for how to act appropriately on such an occasion, but he cautions her not to cry, because that would appear to upstage the family's grief, and they are the stars of that show. The young man's mother earlier had appealed to Jep to help her son, but he backed off, saying he was not a psychologist. In spite of his lecture on funeral decorum, once he is there he seems to be melting (or stripping away his inauthentic layers), and begins to weep spontaneously in a very authentic way, not quite sobbing, but close, as he recognizes his own mortality while grieving that of others.

After this, he and Ramona spend the night together, and when they wake up in the morning, Jep, whose habitual sleeping around no longer fulfills him, says he is glad they did not make love. She replies that it was good that, instead, they loved one another. Through the combined experience of grief and love, he begins to gain or regain an open and brave heart.

Jep demonstrates his Demeter-like kindness as well in his attention to Romano, a male friend who is failing to make it in Rome or with a woman he loves but who only uses him in return. The similarity of the names (Romano and Ramona) suggests a comparable theme of characters psychologically (or physically) stripping down to be emotionally naked and vulnerable. Romano admits to his ordinariness in a public performance and then comes to say good-bye to Jep because he now plans to return to his hometown. Jep learns from this friend, recognizing that he too is ordinary. This begins to free him from his despair and from the egotism that stands in the way of his letting go of his pretenses so that he can be fully present to life. Although Jep was lord of the high life, his focus on being special and the one in charge has kept him from participating fully in the experience of the dance, or even of life.

Jep's Persephone side is awakened further when Ramona dies and he hears of the suicide of his first love, and in a very touching scene, learns from her husband, who read her diary, that Jep was the love of her life, and her husband just a good companion. Often, what stands in the way of any connection with our deeper selves is unresolved grief and remorse. In addition to his grief at these three deaths, Jep feels regret at not doing anything to try to help the young man and an awakened case of sadness about the loss of his first love. All he knows is that she mysteriously left him, and he just let her go, without any awareness that her love for him was as deep as his for her or that he would never experience that depth of love again.

We then relive with him the magic moment in his youth when he bends to kiss her and she turns away, leaving him so momentarily transfixed that he is speechless. Later in the film, we see the continuation of that scene in his memory. She backs off and then turns around, saying that she has something to show him. She unbuttons her blouse, revealing her breasts and looking incredibly goddess-like, then retreats to a cave-like space between large standing rocks on a shoreline. Jep remains motionless, apparently so full of love and awe that he cannot move. We aren't told what happens in their relationship thereafter, since the scene fades out, as was customary in movies in the past when romance is in the air, leaving the viewer to imagine what happened or failed to happen. She is, in memory, his Persephone figure, who as a young woman awakened in him what Jung called his anima, the feminine, soulful side of him.

Jep's early inability to realize the promise of true love is reminiscent of the Henry James novella *The Beast in the Jungle,* about a man, John Marcher, who believes some spectacular fate—positive or negative—is awaiting him. Marcher invites a woman friend, May Bartram, who loves him deeply, to wait with him. Years go by, and eventually she dies. When he visits her tomb, he observes another man, at a nearby grave, sobbing uncontrollably. Comparing himself to the expression of deep grief he sees, he realizes how caught up he has been in his shallow egotism, viewing the woman who loved him only "in the light of her use." Now he recognizes that he missed the great thing that awaited him—which was the opportunity to love her—and he is left with the knowledge that the emptiness of his life was totally unnecessary and the result of his own fatal flaw. This realization is the beast that now springs upon him.

By the end of the film, Jep, who also missed the chance when he had it for a great hieros gamos love, allows himself to feel fully the depth of his love for this woman as he remembers how she had been offering him an experience of great beauty that could have lasted a lifetime. Through this memory, he comes to understand that the only way to escape his sense of alienation is by following a path of genuine passion and love, rather than living only on the surface of the "high life."

But this realization does not fully emerge until Jep feels a spiritual calling to find renewed meaning in his life. Jep's experience of remorse opens him to seek meaning and a sense of calling. Here, the persistent juxtaposition of the images of the beauty of Roman Catholic cathedrals, music, and rituals with those of wild dancing begins to make sense.[1] The archetypal subtext of *The Great Beauty* addresses the dichotomy between the spiritual, in a form that prescribes self-sacrifice in this life so as to have fulfillment in the next, and the profane, which emphasizes dancing and physical beauty.[2] As his spiritual seeking takes root, we again view Jep dancing, this time with renewed joy, commenting that what he loves about the "dance trains" (which are like the American bunny hop) is that they go nowhere, a statement about being completely in the now, without goals—the Dionysian path. Jep is not just having a good but meaningless time. Instead, he is dancing in celebration of, and gratitude for, the beauty of life, and in this way restoring dance as a spiritual practice, which we might see today as a complement to the Buddhist practice of living in the present fostered through meditation.

As Jep and the viewer grapple with the tension in the film between Catholicism and Dionysian revelry, experiences that seem miraculous begin to occur. Jep goes to visit a friend who is a circus magician, and the first thing Jep sees when he arrives is a huge giraffe standing in a courtyard in front of a beautiful, ancient-looking stone wall, illuminated by moonlight. The friend says that he is going to make the giraffe disappear and does. Although the friend then explains that it is just a trick, seeing the giraffe still inspires awe in Jep and in the viewer.

Jep is invited to interview Sister Maria, a Mother Teresa–type nun acclaimed as being saintly for her selfless service to the poor. At 104 years of age, Sister Maria is portrayed initially as almost comically pathetic. She is old and drab, looks like she could fall over at any minute, says hardly a word except a very occasional religious cliché, and leaves her mouth open, showing only two crooked teeth. She clearly never saw a dentist, eats only roots, sleeps on cardboard on the floor, and spends almost all her time helping others. She seems to be a cautionary tale against having too much Demeter archetype in its self-sacrificing mode.

Yet after spending the night on the floor of Jep's room, she goes onto the balcony, where she is surrounded by beautiful, flamingo-like birds that apparently have been attracted there by her. She has read Jep's novel and likes it. Suddenly, she asks him why he never wrote another. Throughout the film, person after person has asked him this same question, and he typically has responded that he cannot do so and party every night, since writing takes peace and time. But to Sister Maria, he answers with more seriousness, explaining that he was looking for "a great beauty" and never found it. She says that she knows the individual names of all the exotic birds around her, then she blows out softly and, in unison, they fly away. Here again, we are witness to a moment of awe-inspiring beauty where it seems as if miracles truly can happen.

With skillful juxtaposition, the film cuts between images of Jep's memory of his first love and Sister Maria on her knees laboriously climbing the Scala Santa, or Holy Stairs (which are said to be those Jesus walked on before the Crucifixion, now in a building at the Vatican). It is painful to watch how difficult it is for her, but it is evident that she persists out of her love for Jesus and faith in God. Her path is to merge completely with God through loving service to others, and thus to transcend this world, while his is to find spirit manifested in it. Hers is inspired by her love for Christ (pictured at the top of the stairs she laboriously climbs), and his by his memory of the flesh and blood woman he loved and the gift of living on this beautiful earth and in a joyous way.

Though Jep's spiritual path is not the same as hers, her example helps him reclaim a sense of purpose at a deeper level than he possessed previously. For any of us, as for Jep, our failure to show up in our deeper and more authentic essence often results from underrating ourselves because we have accepted our inadequacies and ordinariness. However, when we realize that being essentially ordinary simply is the human condition, we can be freed up from our desire to be better than others (for Jep, being the lord of the dance or writing a book even better than his last) to do what is ours to do. At the very end of the film, Jep begins his new novel, for he has discovered his great beauty and has moved beyond egotism to rather humbly pursue his special gift as a writer along with his function as lord of the dance. This suggests that he now has a more developed expression of his Zeus archetypal ally, since he has found the courage to pursue a true gift that requires work—a challenge for anyone with a dominant Dionysian archetype.

Overall, the film's imagery masterfully reconciles the more ascetic spirituality of Sister Maria, which is focused on truths not of this world, with the ability to embrace fully the joys of this world, which is Jep's path. Jep explains the resolution to his existential crisis at the end of the film, so briefly that it is difficult to absorb. He says that some people seek out the beauty in the other world beyond this one, but he is learning to see how that beauty reveals itself to us in this world in between all the "blah, blah, blah." It is up to us to see, hear, and feel this beauty—which is in us and all around us—and to fully treasure being here for the time we have. His final words, "After all . . . it's just a trick," are delivered in the context that he is now able to write his new novel, having found what he was seeking. The "trick," which recalls the miraculous scene with the disappearing giraffe, relates to what we allow ourselves to see and what we don't. Jep has learned how to make life magical by following a more authentic Dionysian path. His decision to write a new novel shows that he has connected with something within that he trusts, so he now knows that he has something worth writing about—the great beauty he always has sought and now has found.[3]

The genius of this film is that it creates a potentially Eleusinian experience—an archetypal field that practically requires us to take in "the great beauty." The power of the visual images and the music abduct us into the realization that beauty is ever present, and the juxtaposition with Sister Maria suggests that this beauty is a transcendent spirit peeking through into secular life in immanent moments that leave Jep awestruck and speechless. By the film's conclusion, we can understand, as Jep clearly does, the quote from Louis-Ferdinand Céline's novel *Journey to the End of the Night* that cowriter and director Paolo Sorrentino uses to begin the movie: "Our own journey is entirely imaginary. That is its strength. It goes from life to death. People, animals, cities, things, all are imagined. It's a novel, simply a fictitious narrative." When we first hear this passage, it sounds despairing. But at the end, we can see that it is about narrative intelligence. The quality of our lives is defined by what we choose to notice and the story we tell ourselves about it. And it is through the imagination that soul and spirit can be touched. That is "the trick" that can help us open to see and feel "the great beauty" in us and around us, so that we can receive the Dionysian gift of ecstasy.

APPLICATION EXERCISE:
Finding the Great Beauty in Your Life

Begin by reflecting on your life, whether it has been long or short, recognizing what you have done that you love to do and whether it still lights you up. Recall times of great beauty and joy, things you feel ashamed of or regretful about, and ego-oriented achievements—anything that might help you find or be in the way of connecting with what is deeper in yourself. Then notice where in your life now you are able to take in "the great beauty" in and around you and who and what in your life helps you to do this. Finally, consider what new thing that beauty might call you to do and how it might affect the way you experience the life you already are living.

Dionysus Lesson Three:

Dancing Collective Joy

P EOPLE TODAY HAVE BECOME more and more isolated from one another. In *Bowling Alone: The Collapse and Revival of American Community,* Harvard political scientist Robert Putnam analyzed data suggesting that Americans were alienated and social community had broken down. It appears to me that this has gotten worse since his book was published in 2000, as people are being stretched in so many directions, the population has become even more mobile, and culture wars and class differences undermine the connections Americans formerly made through things like bowling leagues, the PTA, churches, and various forms of political participation. My late mother lived in Houston, far from her extended family in Chicago, but she had numerous women friends in the neighborhood and in her church. I also moved far from family, but in many of the neighborhoods where I've lived, most people don't have time to get to know one another. More often, my friends have been those with whom I worked, so when I changed jobs, I lost a good many of them.

A major subtheme of this book is the need to be centered in yourself while connected to others. So far, we have looked at this primarily in terms of the mother-child bond and romantic relationships. This section is about learning to feel connected to other people, and not just to those in your own social group, economic class, or vocational field.

In *Dancing in the Streets: A History of Collective Joy,* Barbara Ehrenreich traces the history of people dancing together back to Dionysus, along with antecedents much earlier in prehistory. In fact, archaeologists have concluded, based on evidence from cave art, that collective dancing goes back at least to Paleolithic times and that such dancing occurred throughout the world, sometimes involving one hundred or two hundred people at a time. What interests Ehrenreich is the function of such dancing, why over time it was systematically repressed, and how it can restore a sense of supportive community in our time. She sees the contemporary epidemic of anxiety and depression in the midst of plenty as resulting from this repression, and identifies the remedy in Dionysian energies and practices.

This lesson has both individual and collective importance, showing how being and feeling connected to people in general is an antidote to modern ennui. It also explores how such connectedness helps to create needed social change, as well as to strengthen our bonds to one another through its expression in recreation, family and work life, religion and spirituality, and the broader society. This sense of connectedness is essential to gaining the ability to come together to solve our collective challenges.

Using the archaeological evidence and psychological analysis, Ehrenreich concludes that nature has made dancing in a group enjoyable, just as it made sex pleasurable, so that people will engage in it *for their own good*. In the case of sex, this good is the perpetuation of the species and the human bonding that results in loving care for one another and the next generation. In the case of groups moving in time to rhythm, it fosters group bonding and harmony while also defusing conflict and preventing violence. There is something, she notes, about rhythmic music that inspires the body to move in time to it, whether that is leaping up to join the dance or just tapping one's toes.

Earlier in the book, we examined the role of entrainment in the mother-child bond. Mirror neurons in the brain help a baby form an attachment to its mother, and the mother to her child. These mirror neurons enable the baby to smile back at the mother, make eye contact, and in other ways replicate her sounds and facial expressions. In dance that is performed consciously as part of a group, the mirror neurons of individuals help them connect with the rhythm, the music, and the other dancers, which fosters a kind of group entrainment. We know that group bonding also occurs chemically through pheromones, which align women's menstrual cycles when they live in proximity with one another, and individuals' brain rhythms automatically align with those of others in their surroundings.

When in a group, I often have noticed that I would either be so entrained that I'd just go along with the crowd or fight this by holding the ground of my difference and then feeling alienated and cut off from the others. Through dancing in a group at the time I was studying the Eleusinian Mysteries, I realized that the problem I was having, and one that is common to many people today, was easily solvable. Many of us have learned the important Zeus lesson of establishing clear boundaries so that we neither let others trample us nor conform when we do not want to. However, once we know who we are through our Persephone journey, we are ready to gain the Dionysian secret for being authentic *and* genuinely part of a group.

If we realize that our energy fields already are entwined with others', we can shift the focus from protecting our boundaries to staying centered in our core essence (which can be helped by centering on the vertical midline of our bodies). We then can enjoy the lovely feeling of being ourselves and radically belonging. Dancing in a group helps us learn this skill, since the music aligns the dancers' brain rhythms, the dance steps align their movements, and their bodies' individuality of expression keeps them centered in their uniqueness.

Dancing together was always an important part of women's ancient bonding and their magical mystery traditions, and Dionysian revels reflected a masculine desire to protect this ancient tradition, which Dionysus rites did throughout the Mediterranean. Women's communication styles carry with them, even today, the urge to dance together. Women often face each other, employ hand gestures that relate to each other's, and speak over each other, establishing a bonding rhythm. Experts on communication styles find that this cooperative overlapping demonstrates high involvement in the conversation and with the other person.

Even today, there are many ways we as people "dance" together. If you watch people in a crowd in a place like New York City, you will see them weaving in and out as they go toward their destinations, some rushing and others strolling as if they had all the time in the world. From above, this looks like an intricate dance. To avoid collisions, individuals must be able to focus on where they are going while being keenly aware of all the other bodies near them and shifting slightly this way or that. The next time you are making your way through such a mass of people, apply a Dionysian mind-set to crowds—being very present in your core while also tuned in to the physicality of others—and you will become conscious of how amazing this is and take in the pleasure of "dancing" with everyone there instead of finding crowds stressful.

Journalist David Brooks declares in *The Social Animal* that evidence is all around us that the thinking that motivates most human action is driven largely from the unconscious level and is influenced continually by our surroundings. The conscious mind filters out most sensory input so that we will not be distracted, but the unconscious has a greater ability to process it and directly influence our behaviors. In *The Tipping Point*, Malcolm Gladwell applies this understanding to explain how cleaning up the New York City subways and stopping people from jumping the turnstiles markedly decreased crime. We all, unconsciously, alter our behaviors based on social cues, showing the parts of ourselves that fit in. Psychologist M. J. Apter, in *Motivational Styles in Everyday Life: A Guide to Reversal Theory* and other works, explores how natural it is for any of us to adapt to our environment, without even thinking about it, so that we look serious and studious in a classroom, fun loving at a party or in a bar, reverent at a solemn religious service, or hard working and focused in the workplace.

We are always dancing together unconsciously, which is how several people can cook simultaneously in a small kitchen and not bump into each other, work teams learn to synchronize what they do like jazz improvisation musicians, and most of the time cars careening down a multilane highway do not crash into one another. There is a quality of field sensitivity or diffuse awareness in movement that allows us both to concentrate on where we are going and be mindful of our surroundings.

Many people learn this through team sports, which combine Zeus's focus on winning and Dionysus's on enjoying the game. Players utilize diffuse awareness to simultaneously keep track of where they are in time and space, where the ball and goal lines are, and the shifting trajectory of all the other players. Just as children learn through play, adults often like sports that replicate an archetypal drama. Football, for instance, enacts a war story as a game that focuses mainly on strength and winning. However, American soldiers, especially in the great world wars, were successful, not just because of their strength and courage, but because they had each other's back and could be counted on to try to save their wounded buddies.

Lately, I've noticed this integration of Demeter with Zeus qualities being expressed in media coverage of professional sports. One example: Six games into the 2014 season, the Seattle Seahawks football team, the defending Super Bowl champions, had lost three games. An article by Clare Farnsworth on the team's website summarizes the turnaround that resulted in their winning nine of the final ten games of the regular season, tying the best ten-game finish in franchise history. Although the turnaround by the offense was more traditionally strategic, the transformation in the defense reflected the archetypes in this book. As Farnsworth writes, it began during a soul-searching meeting. According to linebacker K. J. Wright, "Everybody got together and figured out what the issue was and got on the same page. Everybody talked. The players talked. You just felt the team really changed then." The result, he said, was that now it was "just like it's supposed to be. Everything feels normal. Everybody's happy. We're trusting each other. We're loving each other out there. It just feels real good and we've just got to continue it all the way to the end."

Coach Pete Carroll followed up, giving all the credit to the players for accomplishing the "emotional shift" that was needed for the team to succeed.[1] Underneath the sports jargon, we can see Zeus reflected in a collective focused on a win, Demeter and Persephone in the emphasis on emotional honesty and loving one another, and Dionysus in learning to be completely aligned (entrained) in what they were doing as well as being able to experience their work as play. This is a far cry from the traditional macho model in sports and business, and it is a positive sign for male-female partnerships. I've also noticed that a number of men who are full partners with their wives or significant others use team metaphors to describe their efforts to make their families, and the individuals in it, successful.

Baseball provides playful competitiveness, too, but it is configured around a hero or heroine's journey, not war. We can think of the pitcher as fate sending us events (pitches) to which we have to decide how to respond (to swing or not). We then journey (run) around the bases, trying to get to home plate with the help of our teammates, making strategic decisions (steal the base or stay put), hoping not to be tagged out by fate. If we do make it home, the treasure we give our team is a run, and what we give our fans is a shot of vicarious happiness.

There is something about baseball that inspires filmmakers, novelists, and other artists, and that gets the attention of prominent academics. Scholar and poet Barbara Mossberg, a former president of Goddard College, uses baseball to explain American individualism to international visitors. Americans, she tells them, move a lot. We leave home literally and figuratively to seek our fortunes. We have great adventures, and if we are smart and lucky, we find the right home and lifestyles for us. If we combine a can-do spirit with ingenuity and a bit of audacity, it all can be fun.

Anthony Moore, a now-retired administrator at Georgetown University, wrote *Father, Son, and Healing Ghosts* about the movie *Field of Dreams*, interpreting it from a Jungian perspective. He sees the baseball diamond as forming a mandala shape, a symbol of human wholeness, a completion that requires reconciliation with one's origins, so that you come to terms with your biological parent and the parent within yourself. Moore's analysis may explain how the mythic power of *Field of Dreams* is such that more than twenty-five years after the film's debut, thousands of tourists travel each year to the field in Iowa that was the setting for the movie to watch local townspeople reenact the scene of the ghostly players taking to the diamond. I believe the film has had such a lasting impact because, though its story is literally implausible, it is psychologically right on.

In the movie, Iowa farmer Ray Kinsella hears a voice in his head saying, "If you build it, he will come." This propels him to build a baseball diamond in a cornfield; once he does, ghost baseball players emerge from the rows of corn and begin to play. The same voice then sends him on an adventure that eventually allows him to forgive and reconcile with his father, who is one of the ghost players, just as Persephone forgives her father and rejoins the community of Olympian gods. Ray's father wanted him to be a professional baseball player, which was not what Ray wanted for himself. He feels resentful of his father, deprived because he was not that much of a role model, and guilty for the nasty words he threw at him as he left home for good.

Like Ray, we would be wise to let go of any grudges that hold us back from belonging and fully living our lives (which does not mean we have to condone something hurtful done to us), and often life offers us a journey that can heal us if we pay attention. On his journey, Ray connects with his Demeter compassion when the voice tells him to "heal his pain," and he helps a writer he admires recover from disappointment and cynicism. Ray then has an out-of-time experience with a male role model from the past who shows him what a man could be—both in the way he recovers from disappointment in having risen to the majors in baseball but never getting to play and then being the kind of doctor who healed out of a sense of generosity, care, and love. Having become the man he wished his father could have been, Ray can reconcile with his dad in a beautiful scene of wordless male forgiveness and bonding where they play catch on the ball field—a symbolically important gesture, since Ray's rebellion against his father's desire for him took the form of refusing this father-son ritual.

Because Ray has been playing (Dionysus-like) when others think he should have been farming, his farm is threatened with foreclosure. At the film's close, however, a long line of cars, filled with people coming to see the ghostly athletes, can be seen approaching the ball field, signaling a happy ending of restored community and renewed prosperity. On a metaphorical level, this tells us that when we face and come to terms with our resentments and transform guilt into remorse that leads us to new ways of being and living, we can see and open up to new possibilities emerging in our futures.

Sports and sports movies are major ways today that people, and especially men, experience archetypal lessons. Of course, because both team sports and business focus on competition, they reflect an implicit us-them mentality. Such an attitude means that they fall short of what collective dancing, done for the joy of it rather than as a means to an end, can offer in terms of inclusive human entrainment (deep bonding). But learning to entrain with others always begins with smaller groups, and only then extends to humankind more broadly. Further, entrainment in teamwork calls on both Dionysus and Zeus archetypes, potentially integrating them into the group consciousness.

Through bonding by dancing together, Ehrenreich notes, ancient peoples gained courage for the hunt, the collective will to protect themselves from invaders (here also connecting with Zeus as well as Dionysian energy), and group support and clarity of intention as they moved from one season of the year to the next prepared to meet its challenges. Similarly, such dances were conducted for the healing of individuals and to usher in new times of life, such as puberty or midlife, so people did not have to grapple alone with what the change meant (which connected Dionysian with Persephone's energies). In such contexts, the alienation and depression that can accompany such changes, as often is the case today, would be rare.

Ehrenreich traces the beginnings of modern malaise to the loss of rituals of collective joy. She attributes the eradication of such celebrations to the rise of hierarchical and warlike cultures, which emphasized stoic bonding that required leaving one's individuality at the door to follow orders and being tough enough to go into battle. She then shows how ecstatic entrainment (similar to babies' entrainment with mothers and lovers' with one another) historically has been discredited by elite groups that saw it as undermining their power, and that utilized more controlled entrainment through marching and martial music to get people to be willing to fight wars.

To explore further why the ecstatic dance of the Dionysian rites, the Eleusinian Mysteries, and the circle dances of early Christians around the altar are so important, Ehrenreich points to their populist, liberation impulses. These ceremonies typically were open to everyone and attended initially by those from humble backgrounds. She finds the origins of Christianity in this tradition, with the folk hymn "Lord of the Dance" presenting Jesus as dancing his way through death to resurrection and inviting others to follow him in equal fearlessness. Marion Woodman and Elinor Dickson begin a chapter of *Dancing in the Flames* with a quotation from this hymn, observing that it is one of the best-loved of all times. The first verse starts the rollicking song off with Jesus inviting us all to dance with him and goes on to portray him dancing through even the Crucifixion as well as the Resurrection.[2] Although dance eventually was banned in churches for many centuries in Europe, it persisted in festivals and carnivals, but even these were outlawed by the time of the Puritans.

Ehrenreich notes that earlier ecstatic religious forms remain alive to some degree in Pentecostalism but concludes that they are largely absent from organized Christian worship. However, I see the Dionysian alive and vital in many black churches, which are centers for social and political life in Maryland's Prince George's County, where I formerly lived. And if anyone thinks Dionysus is gone from all spiritual rituals, they have never been to a Jewish or Latin American wedding—or the weddings of many other ethnic groups and religious faiths—where typically the community gathers around, dancing in genuine celebration of the loving couple and demonstrating their willingness to support the couple's commitments and vows.

Some progressive and ecumenical movements are reintroducing Dionysian elements to bring the young back to connect the young with the divine. The Creation Spirituality movement uses light shows and imagery, along with music, to reintroduce ecstatic dance as a form of worship to augment traditional spiritual practices, including a cosmic mass to celebrate the universe as God's creation with a light show, images flashing on the wall, and danceable music.[3]

At my local YMCA, I found Nia, which is a dance form designed to foster the ability to move with pure joy and also to promote the concept of dancing throughout life. What this offers is the capacity to feel pleasure in everything you do through being in touch with bodily sensations and your energy field.[4] The 5Rhythms dance form, based on the work of Gabrielle Roth, helps people "sweat their prayers" by moving through the core rhythms of life.[5] Because it provides catharsis and then stills the mind, this practice also offers some of the benefits of meditation, as does Nia, with its focus on being fully present, always sensing your body.

Ehrenreich observes that although sports fans for the most part are spectators, they share a strong identification with their group through allegiance to the same team. It does appear that this serves some of the purpose of group bonding that religions traditionally have provided, although it remains in an us-them context. In older stadiums, people of all classes sat side by side, but this democratic feature is absent from new arenas, with their special sections for the elite. Spontaneous synchronized movements, such as the wave, are populist in nature, and players themselves frequently do victory dances when they score or win. An even stronger example of the return of Dionysian energy is found in large rock concerts or in small club venues, where the star is the stand-in for Dionysus, and where individuals dance in their own ways but bond by moving to the same rhythm and participating in an experience they love. Carnival in Brazil and Mardi Gras in New Orleans maintain this ancient feel, as masses of people flock to the street to dance.

Most telling, however, is the way that liberation movements incorporate music and dancing when they gather, and how they do so on public greens and in parks, which often is the closest we come to the natural settings where Dionysian rites took place. Ehrenreich emphasizes how often music and dance accompany protest demonstrations around the world, especially among the young. Both the Eleusinian Mysteries and the overlapping Dionysian revels were open to all—men and women, slave and free—and thus themselves were liberation movements in a class-based society.

In Santa Barbara, California, where I lived while writing this book, Dance Tribe is an organization that holds weekly ecstatic dances and sees dancing as a way to commune with ourselves and with each other, and throughout the world there is a growing movement to offer alcohol- and drug-free venues for dance as a spiritual practice requiring no particular belief system. What is spiritual about it? The experience of everyone dancing to the same music, but each expressing his or her own authentic self *and* entraining with the group, can foster communal joy. As we have seen in Jep's story, there is a vertical and horizontal side to spirituality. The vertical pulls attention up to contemplate the divine, and the horizontal expands us out to love one another and the earth—hence the archetypal symbol of the cross.[6]

Today, the intent of Dionysian maenads coming to town and playing music to invite spontaneity and greater aliveness is at the root of flash mobs, where secretly organized individuals emerge in a public place singing or dancing, often getting the entire crowd to join in. This is now a huge movement, and you need to see it to believe it.[7] You also can express your Dionysian spirit any time you, alone or with others, surprise people by getting them singing and/or dancing. When I was the provost at Pacifica Graduate Institute, a group of faculty members from our Counseling Program, having thoughtfully checked with me, interrupted what usually was a very serious academic meeting to jump up and sing a rap song in honor of a retiring faculty member, and of course, everyone joined in. This kind of informal flash-mob activity is happening all over, and some people have a knack for being a flash mob of one, doing or saying something that makes a meeting or party suddenly come alive.

Joining the dance can be interpreted metaphorically as getting out of the spectator mind-set, where we criticize as if we were on the sidelines of life. Instead, we can become person-positive viruses, spreading aliveness and joy wherever we go or to whomever we are with. This might include a spontaneous random act of kindness, smiling at someone who looks lonely, taking a moment to chat with people you meet, or walking with a skip in your step. It can be putting aside your list of things to do to follow an impulse without worrying about what people will think. My friend Barbara Mossberg once surprised a group I was in by forming us in a circle and getting us to do the hokey pokey, putting first one arm in, then the other, one leg in, and then the other. Finally, with great aplomb, she got us to dance putting our whole selves in, proclaiming, "That's what it's all about!"—meaning *life*. We evoke the archetype of Dionysus when we put our whole and real selves into everything we do, and it is even better when a group joins in with us.

Lately, I have seen many articles bemoaning the younger generation's addiction to being in continual virtual communication. Yet all over the world, the young have self-organized rallies and even revolutions using Twitter and Facebook, as in the Iranian Green Revolution or Arab Spring, standing up against tyranny and all kinds of injustice in service of a better world. While such massive events have not always led to optimal outcomes, they suggest that youth today are learning to entrain virtually in the service of the call of the future, something any of us can do. People of all ages are now posting evocative images and stories, and increasingly videos, responding not only to the content of others' posts but also to the rhythm of their frequency and energy. This can be experienced as a kind of virtual dance-like interaction. When you use the medium that way, you might find not only that you are happier but also that you are spreading happiness that can go viral. Who knows how many people will be cheerier because of you?

Global virtual interconnectivity is allowing people from various cultures to feel as though they are a part of the same dance, even when they are at a great distance from one another. YouTube and other social media sites carry videos of groups dancing for peace or to end violence against women or for other causes. These videos are widely available around the world. Sometimes they can be synchronized with sites in different time zones to give the sense that we all are dancing together.

However we find ways in this new world to entrain with one another, it is important that we do so. Ehrenreich concludes,

> The capacity for collective joy is encoded into us almost as deeply as the capacity for the erotic love of one human for another. We can live without it, as most of us do, but only at the risk of succumbing to the solitary nightmare of depression. Why not reclaim our instinctively human heritage as creatures that can generate their own ecstatic pleasures out of music, color, feasting, and dance?[8]

The more that we do the things that have always helped people love their lives, the less depressed we will feel and the more we will be able to work together to solve the issues before us—full of optimism, as happy people tend to be.

APPLICATION EXERCISE:
Dancing Through Life

Experiment with finding ways to experience entrainment with a group as you go through your day. A first step is to get centered in yourself using earlier exercises provided. The second is to relax your body so that it is not defended against others or against life. You can do this by imagining the open feeling in your body when you are welcoming someone you love and feel safe with and close to. Then see if you can hold on to this in interaction. One way, of course, is to go dancing and remain very conscious of moving in your own fashion and being conscious of others and allowing their movements to naturally affect yours. As you walk around town, loosen your body so that you are sauntering a bit and open up to feeling your connection with others around you, weaving in and out in a normal way, but with consciousness of this being a dance. If you have time, you also might let things that grab your attention take you off course as you move toward them, admire them, and then circle back to your walking. You can even dance with the weather, responding to the pleasure of a soft wind or enjoying a quick sprint to get out of the rain. You can bring this consciousness to driving on a highway, enjoying dancing with the other cars in the safest manner possible. Once you have this down, you can practice entraining with a group in your family, a friendship group, workplace teams, or any other groups that matter to you, in person or online.

Dionysus Lesson Four:

Directing Your
Inner Theater Company

THIS FINAL LESSON EXPLORES the invention of drama and the related psychological tasks of learning to write your life script and call on parts of you as your inner actors, so that you can thrive in the improvisational theater of life.

It is difficult to imagine a world without plays, films, and television dramas. Yet there once was such a time. Drama, at least in the West, originated in Athens with ancient hymns sung to honor Dionysus, in which various individuals would sing different parts. In the sixth century BC, the Dionysia (a festival held in honor of Dionysus) included competitive music, dancing, and poetry. Out of this tradition came the birth of drama as we know it today, including both tragedies and comedies, both of which mirror issues in the culture of their time.

The tragedies warned of counterproductive behaviors, like hubris, so that leaders and others would not fall prey to them, and also provided an opportunity for the collective catharsis of deep emotion. The comedies typically satirized issues of the period, with emotion released through laughter, encouraging group happiness. The most famous of the ancient comedies is Aristophanes's *Lysistrata,* which presents the women of Greece as antiwar activists, refusing to have sex with their husbands until they agree to make peace and end the Peloponnesian War.

In classical Greek times, the chorus spoke lines in unison, reflecting the community's views about the goings-on in the play. Like the inclusive Dionysian dances, the plays had a populist intent, allowing the opinions of the common folk to be shared with the elites, all of whom would be present for a major theatrical performance along with their households, including servants and slaves. Theater, at its best, has always mirrored what is going on in society and generates discussions that can lead to social or political change. In Greece, plays about the hubris of kings were meant to warn the actual leaders of the city not to fall into that trap, and to predict the downfall they would experience if they did.

Identifying with a character, as the kings ideally would do, can help us observe ourselves and give us feedback so that we can correct our dysfunctions and get back to enjoying life. But figuring out what to do with mirrored feedback can be complicated. If others are identifying with the same characters that we are, talking with them about those characters can help us think through what it is we need to learn without having to reveal our, or their, private issues. A character in a play provides a third thing that can be discussed in ways that get both of you, or a group of you, out of yourselves so that you can see your own issues from a different perspective—often one that is archetypal.

It is wise to be alert to mirrored feedback in life in how others respond to us. Let's suppose that you are saying or doing something and people ignore you. Sometimes this tells you that you are off base and need to change, other times that what you are doing or saying is inappropriate to a time or place. Even if your behavior or what you say is not necessarily wrong, the feedback can let you know that it just isn't a good enough match with the mind-sets and stories being lived out by those around you, so to be heard and noticed, you might want to figure out what story they are living and how you could communicate using its language. It can be helpful to imagine that you have stepped onto a movie set where they are filming a particular plot in which what you are saying and doing does not fit naturally, so it disrupts the flow in progress. What to do? Perhaps you were bringing a caring message and then realize that you stepped into a war story. You then can recall that caregivers get noticed and appreciated when they start helping the wounded or feeding the troops. In this case, you might consider who is feeling wounded or hungry for sustenance and how your message can be tweaked so that it is recognized as useful to those involved in the situation they are facing.

Greek theater provided a unifying, shared emotional experience, as did Dionysian dancing. Although the audience at a play is not moving together, they all are experiencing the same story, with similar attendant emotions. The end of Jane Wagner's *The Search for Signs of Intelligent Life in the Universe* helps us recognize how very special that is. In this one-woman play (performed by Lily Tomlin), Trudy, the wonderfully wise but crazy bag lady, is trying to help aliens understand our world. They are particularly stumped by two questions: First, what are goose bumps? And second, why is a can of Campbell's soup just soup, while an Andy Warhol painting of the can is art?

At one point, Trudy is standing with the aliens at the back of a theater, during a performance, when one of the aliens suddenly shows his arm—with goose bumps. Trudy tells us, "I'd forgot to tell them to watch the play; they'd been watching the audience! Yeah, to see a group of people sitting together in the dark, laughing and crying at the same things . . . well that just knocked 'em out! They said, 'Trudy, the play was soup, the audience, art.'"[1] At any great play, we, in the audience, are moved together, especially if we are so full of awe that for a moment we cannot even cheer or clap.

We often forget how good it feels when we are fully connected with a group and being un-self-consciously ourselves. This can occur when we share delight at a wedding or watch a toddler learn to walk, or do something together that we all love to do. When managing a family meal, we can remember Mrs. Ramsay (in *To the Lighthouse*) creating magical moments at the dinner table, where people's egos fell away, and suddenly they were a community together. In the theater of our lives, there are abundant ways we can promote such times of shared experience, which can happen when people are confronting tragic deaths and feeling the comfort of being moved together, not alone. The satisfactions people gain from shared experiences are an advantage to seeing some films and dramas in the theater rather than at home on a TV screen or monitor—never mind on a tablet or cell phone.

In contrast to storytelling, in which there is one narrator, theater employs multiple actors who play different roles, underscoring the subjectivity of opinion and fostering the audience's ability to understand various points of view. At the end of a play, the audience would want the characters to work things out, so that they could leave the theater feeling cheerful rather than dispirited. A bit of thinking about this makes it clear that the consciousness of theater is related to that of democracy, in which resolving various interests always is in play. Thus, it is not surprising that theater and democracy were invented by the same small group of people in one township (Athens) that also observed the Eleusinian Mysteries as its most cherished rite. Manly Hall, referenced in the Introduction, described how the Demeter and Persephone story and the wisdom of the Eleusinian tradition influenced Freemasonry, America's founders, and, through them, the Declaration of Independence and the Constitution.

This entire book is about the power of story. It encourages you to exercise narrative intelligence so that you can learn from the wiser stories available to us, rather than having to learn everything the difficult way through the school of hard knocks. Imagining your inner life as your own Greek theater can be a means to enhance your narrative intelligence skills. You can begin by asking, "What story am I in?" and then consider what archetypal narrative is playing out and what it requires of you. Your archetypes would be the actors in your theater company, so you can recognize whether you need to call up your Zeus, Demeter, Persephone, or Dionysus (or some other archetype) to respond to a situation in which you find yourself.

Various archetypes want and like different things, which means that organic work-life balance comes from matching archetypal motivations with outer roles and activities. You can recast your inner actors into the roles they would be best at playing and give them pointers on how they are doing. Many people feel burned out and unhappy, not so much because they are working too hard, but because they are depleted by what they are *not* doing. Their real desires are getting no play. It is exhausting to repress all or most of yourself.

If you start to feel weighed down with responsibilities, it can be a warning signal that you are cramming into your life too many extraneous activities, that you are trying to do everything yourself when you need to ask for help, or that one or more of the archetypes is trying to take you over and needs to be reined in. For example, Zeus would have you win at everything, but maybe you don't need to win at beach volleyball on your holiday to have fun. Left to themselves, each of the archetypes may aspire to control your life and generate thoughts criticizing or even shaming you for not having achieved enough (Zeus), for not helping everyone who needs it (Demeter), for not being in touch enough with your deeper self (Persephone), or for feeling down when your life is just fine (Dionysus).

Each of these archetypes also has traits that can help you defend yourself from this archetypal onslaught and the resulting sensation of being overwhelmed. Zeus can explain in no uncertain terms what the roles of your archetypes should and should not be for you to maintain a peaceful and prosperous kingdom. Demeter can educate these inner allies on how they can help you care for your own needs. Persephone can create a meaningful ritual action, like eating the pomegranate seeds, that embeds a decision in your unconscious mind. And Dionysus can shed useless layers, dying to inauthentic ideas, roles, and behaviors in order to make room for fuller expressions of your optimal self.

Lacking connection to any one of the four Eleusinian archetypal stories can hamper success. However smart you are, if you don't have access to your power, you may not be able to act effectively on what you know; however powerful you are, if you are not connected to your heart and soul, your actions can be harmful to you and others; however deep and soulful you are, you cannot fulfill your material needs without the power to act; and finally, however powerful, loving, and intuitive you may be, you may lack the ability to enjoy what you have.

Employing multiple lenses through which to interpret events can prevent any one story from locking you into a position you later regret. In *Thrive: The Third Metric to Redefining Success and Creating a Life of Well-Being, Wisdom, and Wonder,* Arianna Huffington confesses how she internalized an attitude from the culture around her that working around the clock in the service of money and power was evidence of being a successful and admirable person. But when she did this for too long, it led to a physical collapse. She then extrapolated from her experience to warn other overworked people of the impending tragic outcome of a counterproductive story. The problem she exposes is a cultural out-of-control Zeus archetype, not balanced by Demeter, Persephone, or Dionysus. It is difficult not to be affected by such a societal archetypal possession. I find it helpful to remember that this is just our culture's obsession at the moment and does not have to be mine—or yours.

Most of us have archetypes that are the stars of our show, but if we let ourselves become one-dimensional, we become walking stereotypes of an archetype in its counterproductive form (as above with Zeus). People under the influence of Demeter can burn out if they cannot stop doing for others as long as anyone needs anything—and there will always be someone in need. Allowing more of our subpersonalities, who are inner characters of our dramas, to shine provides us with a greater range of abilities and helps us to feel more alive. Plus, if we have some archetypes that never get on stage, they may get antsy and try to sneak in, generally causing Freudian slips and embarrassing incidents where you act in ways that do not seem at all like you, and you wonder, "*Who* did that?" or think, "The devil made me do it!"[2]

Our Dionysian parts love to live spontaneously, which means that they like the improvisational theater of life, where we are faced with new situations that require a quick response. When something surprising happens, we do not always have time for our cerebral cortexes to work through what stories would be optimal to think, tell, and live. This is where having lived these four archetypal stories becomes most essential, so that you have some range in your spontaneous responses rather than relying only on your behavioral habits and your default stories, meaning those you have on automatic pilot. For example, most of us turn immediately to unconscious narratives that see us as victimized, trapped, or mistreated. Such stories trigger an instinctual fight, flight, or freeze response.

Most of us also have automatic responses that began with trauma in some form, originating in occurrences such as being ridiculed, shamed, or mistreated as a child or witnessing something painful happening to someone else. Watching your thoughts while meditating or engaging in mindfulness while going through your day can alert you to these default stories and behaviors. Examples of such stories include thinking of yourself as the heroine of a narrative structured around a plot that continually plays out the experience of being unable to find success or happiness—because your dad never did, because you are shy or intrinsically flawed, or because nothing ever works out anyway. The only sure way to rid yourself of these kinds of stories is to replace them with healthier and more empowering ones.

The teachings and practices of many religions and philosophies support this process. In the Buddhist tradition, you are encouraged to practice welcoming whatever comes in life with loving-kindness. If you do this long enough, a loving narrative can become your default template, with variable details, depending on what transpires. This archetypal story then will determine how you act without your having to think about it. Thus, in your story, you would always take the role of a loving, accepting person responding to whatever comes with patience, love, and kindness. This story, of course, would have infinite variations.

It takes some discipline to shift the scripts you have on automatic pilot, so it helps to have group support. But with or without such support, if you persist in eliminating narratives and behaviors that reflect the old plotline and in living in keeping with the new, your story and your life will change. As we have seen throughout this book, we cannot stop making meaning in stories, but we can rewrite a story that has been replaying itself for most of our lives. When we become conscious of our inner stories, we might think, "Wait a minute, that's not my story. It is one written by my mother, my aunt, my mentor, my minister, my spouse, or someone else—not me." No longer allowing others to be the authors of our individual life dramas is a huge break-through that can change everything. While we cannot always change what happens to us, we can use our narrative intelligence to choose what story we tell ourselves and others about it, just as Zeus shifts his story from control to collaboration; Demeter, from powerlessness to power; and Persephone from being an abductee to realizing her destiny. Similarly, we can identify what stories we are telling that are making us unhappy and limiting our options and then explore alter-natives that open possibilities.

Remember that archetypes, when awakened in you, flesh out your capabilities; as you integrate their strengths in particular aspects of your psyche, they begin to feel like part of you. It is wise to educate and direct your archetypes, asking them to be your allies in learn-ing to master certain activities and not others. Then you can call up the subpersonality that has gained mastery in an activity through the influence of the archetype to help you with a looming challenge. Of course, what you really are asking to appear is a set of abilities the archetype has helped you gain. In the same way, you can be the direc-tor of your own play by educating your various inner archetypal stars, supporting characters, and walk-on parts about how they can contrib-ute to this new and improved script. This is important, because you are not going to win a marathon if your inner Demeter takes charge and is too busy making sure that all the other runners are safe to get you across the finish line; and neither will you have fun at a party if Zeus takes over and starts trying to control who does what when.

Moreover, we can recognize that while we have multiple actors within ourselves, what the world sees is someone performing a one-person show that is so authentic and real, it helps others connect with their own deeper human qualities. In this way, you can have the kind of positive effect that great works of art have. When you are unlearn-ing the negative side of an archetype, you can coach your body in method acting to find and live out its more positive aspect until that becomes habitual for you. And when you are feeling overwhelmed or stressed, you can use your archetypal awareness to access the peaceful part of you.

The Eleusinian rites, as the self-help movement of their time, always promised happy endings, and the narratives of each of the four archetypes in this book have provided models for attaining them. Jung saw the mandala shape—with a clear center and then four parts—as a symbol of wholeness, and many of his patients spontaneously painted or drew such shapes when they were near completion of their individuation process. Mandalas are a bit like a Native American medicine wheel, where the four parts are associated with the four directions— north, south, east, and west—and usually with elements such as air, water, fire, and earth, and sometimes animal totems as well. All these symbolize natural and human archetypal patterns. You might imagine yourself at the center of such a magic circle, with the four archetypes highlighted in this book around you, so that you can call on them when you need them.

So often, people today end up living a story no one wanted, especially when country, company, or family politics heat up, just because they do not know what else to do. Frequently, they start fighting when they actually would rather do something fun, or at least productive. When this happens, it is a good time to call on one or more of your archetypal allies to help you shift the story. This can be a bit like improvisational theater, where unexpected, surprising actions or dialogue change the direction of the plot and then the other players are challenged to react spontaneously.

There is a rhythm to life, and if we stay connected and utilize our full capacities, we often can do just the right thing or stop doing the wrong thing, even though we don't know why. In Thomas Pynchon's novel *The Crying of Lot 49*, the heroine, Oedipa Maas, walks into a ballroom where deaf mutes are dancing in coordinated fashion to music they cannot hear, and all stop when the music does. The narrator refers to this as the anarchist miracle: when things occur that seem impossible. However, this "miracle" actually is quite possible, since the deaf can feel vibrations. As part of the improvised worldwide performance that is always taking place, we can participate in the ongoing cultural conversations happening at a micro level in hair salons and coffee shops, on public transportation, between neighbors, and so on, modeling genuine listening as well as sharing our own points of view.

Dionysian entrainment can be utilized in conversations when our body signals reflect true regard and respectful listening and speaking while matching the rhythm of the discussion in at least the beginning of what we say, shifting gradually if we want to energize or calm down a room. Such dialogue can lead to higher-order interpretations that, over time, could move from multiple "my stories" and "your stories" toward a collective "our story." That is what informs positive collective action. The civil rights movement and the women's movement both started that way, and the same sorts of conversations are going on now. Virtual technologies also can expand our ability to engage in cross group dialogue to achieve cultural consensus. While this lacks the intimacy of physical proximity, it adds the benefit of greater comfort with telling it as it is (at least in the teller's view). The drama that we enact collectively can realize the promise of government of, by, and for the people—the dream that began in ancient Athens and has evolved in its expression over time.

For this reason, and more generally to aid you in applying what you have learned in this book, the conclusion that follows presents additional heroines' tools of narrative intelligence so that you can consistently enter collective dialogues with communication mastery. Of course, the story you tell also is entwined with those of all the people you love and work with, as well as with others important to you, resulting in interlocking archetypal narratives that may mesh or conflict. You might host conversations that share archetypal concepts and vocabulary with family, friends, and loved ones to promote a new level of communication between and among you, and to help you articulate your views as being those of an archetype, which makes the debate less personal. When we identify our differences as between, say, Demeter and Zeus, it becomes easier to find an overarching "our story." If your consensual story is a version of what you have known before, it can feel comfortably familiar. If it is a new archetype, or a version of one with a plot twist, it can feel both disquieting and enlivening because it makes us stop, think, and perhaps act in new ways.

Being conscious that you, as an individual, are embodying the great archetypal stories can remind you to live them in their grandest and most beautiful forms, inspiring you to greatness as an awakened heroine. As you express your special gifts through the archetypes that are active within you, you will reinforce the better sides of those same archetypes in the lives you touch, and in the process, draw people, success, prosperity, and love to yourself. But this is not an invitation to grandiosity. Your task and mine is to be true to our natures and do our parts, trusting that our subplots are one fragment of an unfolding story much bigger than we are, and that we all matter to those around us and to the fate of the world.

~~~~~~

APPLICATION EXERCISE:
Creating Your Own Playbook

Prepare for an upcoming challenge by creating your own playbook. What is the setting? If it were the beginning of a drama, what kind of story would it likely be? Who are the other characters involved besides you? What are the archetypes at play in the characters, and in you, related to this situation? Given what you know now, what is the likely plot structure that naturally would unfold, given the above? That is, who is likely to do what, how will others react, what situation will then be created, what is the likely challenging climax and outcome?

As the heroine of this story, what might be required of you to move all this toward a happy resolution? If you find there are gaps in your ability to play the required part, what archetype's lessons might help you develop a needed subpersonality or skill? As the writer of your play, draft a quick synopsis of the desired plot, and then, as director, instruct your inner characters on what they should be prepared to do. As the actress playing you, work with method acting to prepare to call up what you need as the play progresses. In real life, you will not be able to control what the other people involved as characters in your drama do, so think of this as improvisational theater, and be prepared with as many strategies as you can think of to respond to twists and turns in the plot as they occur. Then imagine your play in the theater of your mind's eye, concluding with a standing ovation as a tribute to the truth and impact of your performance.

CAPSTONE EXERCISE:
Dialogue with Dionysus

In a journal or notebook, begin a dialogue with Dionysus where you speak as yourself and the Dionysus of your imagination speaks for him. Start by thanking Dionysus for how he has enhanced your life; then ask how he would like to be expressed in you. You can then say what you would like from him, including where you think he would be helpful and where not. Allow a natural conversation to develop where you come to an agreement about Dionysus's future role in your life. End on a note of appreciation.

INTEGRATIVE CAPSTONE EXERCISE:
Your Personal Eleusinian Mandala

Create a mandala with symbols, images, and/or words in each of the four corners that represent Demeter, Zeus, Persephone, and Dionysus in your life right now (or how you want them to be). Then put a picture of yourself in the center. This can be your ongoing reminder of what you learned by going through a virtual Eleusinian initiation in the form of this book.

Conclusion

~~~~~~

The Power of Story
to Transform Your Life

B Y NOW, IT MUST BE CLEAR TO YOU that if you want to change
your life, much less your relationships, your family, your workplace, or the world, you have to change the stories that are fueling
whatever drama is currently unfolding. Experts in artificial intelligence
coined the term "narrative intelligence" related to their aspiration to
teach machines to generate and tell stories, since this ability is recognized as the most human of tasks, one that integrates thought, feeling,
and creativity. And because communication is a female strength, this
final chapter is designed to help you hone your narrative intelligence so
that you can be ever more effective in realizing your potential.

Heroines' Tools

Persephone Rising began by stressing the power of archetypal and
mythic narratives, and such stories have been emphasized throughout

the book. However, narratives need not necessarily be archetypal or mythic to have an impact, and the book describes numerous instances when characters improved their situations by questioning stories that had power over them without being eternal narratives. For example, Demeter lets go of the story that she should be a nanny, replacing it with, "I need a temple, and you will build it for me."

In our lives more generally, some stories seem just to be entertaining, but even then they can spark the imagination, and many have a moral, gently conveyed. Some are anecdotes about history or the world around us, or visions of the future, while others may have the same kind of content but are more formal creations with clear plotlines. As we have seen, some stories carry expectations of the culture or the group, and some reflect your individual experience and imaginative flights of fancy. These concluding pages equip you to employ what you have learned, not only related to the four archetypal stories within a powerful mystery tradition, but also in terms of narrative savvy more generally.

In mythic stories, heroes often are outfitted with weapons, like swords and shields, to win a conflict, or if their task is more magical, with wands or grails to transform the situation. More often today, men and women change events through communication, and very specifically through the stories we tell, whether they are archetypal, mythic, or more personal or particular to a situation. Most of us hope that more wars can be avoided and peace talks can be more successful as people get better and better at addressing conflict with words, not fists or weapons.

However much we, as women, may get teased for talking endlessly with our friends about the issues in our lives, communication continues to provide a feminine advantage—in thinking questions through without leaving out their emotional dimensions, in anticipating what others might do in a similar circumstance, in confronting problems we have with others without unnecessarily hurting their feelings, or in expressing what we want and need in ways that can be heard. Narrative intelligence is a critical skill because it enhances strengths that are associated with success for anyone, but particularly for women.

Let's now look at women's communication tools that serve purposes similar to those of mythical swords and shields, grails and treasures, and magic wands.

The Sword and Shield of Story Vigilance

Because the stories we tell about what happens around us are filtered through the archetypal and other lenses in our psyches, they have a subjective element to them. Therefore, it pays to wield the shield of story vigilance, to sort out what is true and what is not—not only in what others tell us, but also in what we tell ourselves. The stories we tell about experiences we have or witness reflect the archetypal configurations of our thought patterns. In that way, what we take to be reality is shaped by who we are and what we already believe. That is one reason the narratives we tell about what is happening around and to us differ so greatly. Much of gender bias happens because the powers that be (who still remain predominantly male) are consciously or unconsciously convinced that their approaches and standards are correct because they are their norm. As a result, they may view what many women say and do as wrong and in need of correction rather than as different and potentially useful in how they could expand thinking and the options available.

And of course, stories also *can* be consciously or unconsciously employed to manipulate people and shape their views around any issue or concern. Enjoying a story requires momentarily imagining it as if it were true, but then it is important to view it critically, especially if the moral of the story being told pressures you to give someone your money, your hard work, your vote, or your heart. Sometimes such stories are based on partial truths, leaving out aspects that do not support what someone wants you to think or not mentioning what is in the fine print.

Your inner sword of critical thinking can cut through lies, distortions, manipulations, and simple ignorance masquerading as truth, working with the shield of your discernment to determine which narratives are helpful to you and which are disempowering or just irrelevant. You began internalizing stories when you were young—not just books read to you or lectures about how you should and should not behave, but also anecdotes about Aunt Janet or the president. As a child, without the capacity to challenge them, you would have internalized these stories innocently, as we all have.

Most of us do not even notice how much, even as adults, we accept the ideas implicit in the stories we encounter on the news, in the movies, on the Web, on talk radio, or in social interactions. This is especially true when those around us act as though the stories they tell are absolute truth, not just their interpretations of events. As an ongoing practice, story vigilance can shield you from internalizing narratives that could disempower you. To practice such vigilance, it is helpful to become increasingly aware of the invasive plotlines that already have found their way into your psyche's garden. Only when you weed that garden can your authentic and beautiful flowers flourish.

Let's take a few contemporary examples of stories that make women's lives more complicated. Many aspects of modern life that stress everyone, but are particularly difficult for those already pressed for time, are just stories, not inevitabilities. Advances in technology have changed our lives in ways that seem magical and that allow so many things to be done more simply, and most of us are able to stay connected with those we love and with people all over the world. However, a story has emerged that such technology means we must be available all the time—to colleagues at work, relatives, friends, you name it. This narrative, not the technology itself, has led many people to, in effect, never be away from their jobs.

Moreover, when the office was largely a male preserve, after the establishment of the hard-won forty-hour workweek and before sex roles became more fluid, few employees worked as many hours as they do now.[1] Business experts some years ago created and sold a narrative that greater efficiency could be achieved (while increasing profit margins or mission accomplishment) partly through cutting staff, especially middle managers. This has resulted in professional and managerial employees of both sexes working harder and longer, often to the detriment of family life, with a narrowed pipeline to advancement and potentially a less resourceful, less efficient work-force.

In this land of plenty, so many of us—rich, poor, and somewhere in between—feel overwhelmed and exhausted from running ourselves ragged. And although we see lots of media coverage of the plight of elite women climbing the corporate ladder, the reality is that they at least have the resources to hire help. Women without such resources typically are stretched much thinner, with some working two jobs if they can find them just to meet basic needs as growing economic inequality expands this divide.

At the same time, new stories about what women should do are emerging that up the pressure on mothers. Homemakers used to send their children out to play, while they had coffee with their friends or even played bridge. Now, both stay-at-home moms and working-outside-the-home moms (and increasingly dads) feel as though they need to be chauffeuring kids from one developmental activity to the next while scrapbooking, Facebooking, and Tweeting every moment in their kids' and family's lives.

So if you find that you are exhausted all the time, you might want to pay attention to what narratives are driving your behavior and determine how well they correspond to your external situation and your real wishes and desires. Would you lose the love of your fam-ily if you were not supermom and superhomemaker? Would you lose your job if you worked at a sustainable pace for sustainable hours? Is everyone you know who has a happy family and a secure job working as hard as you are? If you are working harder to buy your kids name-brand clothing and shoes or your family a minimansion, can you ques-tion where the story came from that you should push yourself to do so? And is what you are doing making you and those you love happy?

Standards of fitness and beauty also have escalated. If, after reading women's magazines, you think that you will never marry or that your husband or partner will leave you if you are not beautiful enough, take a walk through the shopping mall and see that not all couples who are having a good time together meet these unreasonable standards. Even people out there who are trying to help us can overwhelm us. In the period when I'm writing this, I and so many other people I know are, say, trying to fit more time into the week for meditation, to walk and keep track of ten thousand steps every day, to remember to think positive, Tweet, cut out sugar, and so on. The list just keeps going. All the self-improvement strategies available today, however useful they might be, can add to the belief that we can never do enough.

Prevailing stories reinforce the idea that if we cannot keep up with unrealistic expectations, there must be something inadequate about us. So if we cannot "do it all," we are not up to the job, and if we are poor or otherwise struggling, it is because we are not smart or hard-working enough. It takes a powerful narrative-savvy sword and shield not to internalize these stories and begin to tell them to ourselves, with the result that our internal narratives are always pushing us to keep up and chiding us if we can't.

If you are feeling that *you* are not enough, pay attention to where that idea came from and weed your internal garden by choosing to control your own story. It also might help to remind yourself that you are the only you in the world and that you are here for a reason, and to ask yourself, "What in all of this is truly mine to do?"

Your Grail: Finding Meaning and Your Authentic Story

Stories are powerful containers for values and insights, and in this way, they serve as the grail (treasure) that the heroine finds so that she can transform her world. By recounting a vibrant personal history or describing significant moments of development and change, your life stories capture meaningful insights that can help you understand who you are. We see this in accounts of the founding of a country or a company or the history of the ups and downs in a family, and we can imagine ancients enjoying the Demeter and Persephone myth as an origin story for the Mysteries that helped them. In healthy organizations with high morale, employees light up when they tell about how they came to be hired and what it meant to them. Relating the stories reinforces the sense that what they are committing their life energy to matters. Couples never seem to get bored retelling the story of how they met, since those accounts reinforce their love for one another. You may find that some stories you think or tell about your life give you energy because they reveal something real and genuine about identity and perhaps even give hints about your life purpose.

Narrative therapy is now an established psychological approach, and doctors are adopting strategies of narrative medicine because the bigger picture of a symptom typically is embedded in the meaning a person makes of it. Marketers used to focus their efforts on describing the attributes of a product or business, but now they know that nothing establishes a strong brand identity better than association with a compelling tale that communicates the meaning and values of the brand. Thus, many ads now take the form of a story designed to tell as much about the values of a company as the virtue of its products. Who, once they knew to think about it, would miss the archetypal narratives at the root of Hallmark's love-story commercials, which urge us to buy its cards to tell people we love or care about them; Harley-Davidson's outlaw image, offering people (often doctors, lawyers, and other professionals) the chance to get on a bike and feel free; or Dove's regular-gal focus on the beauty in every woman?[2]

While we can have a visceral experience without a story, making meaning of that experience requires a narrative. For example, you come home from a date with a very nice person, which is an experience. Especially if you are a woman, you begin to spin out a narrative that makes meaning of the event: "We had a good enough time, and it was okay, but I don't want to do that again, so if he (or she) calls again, I will _____." Or "That was amazing, and I think I'm falling in love," which can be followed by a fantasy of wedding bells or any number of other stories that immediately might come to mind. Your boss snarls at you—which is an experience—but then you think about what it means, and that leads to constructing an imaginary plotline of how you will try harder, or of what you will do if you are fired, or some such thing.

Some hard-won meanings can save your sanity or your life. Viktor Frankl, in *Man's Search for Meaning*, discovered that people were more likely to survive being imprisoned in a concentration camp if they created a narrative that enabled them to find some positive meaning in their experience and thus a reason to live. The narrative could be about seeing one's loved ones again, dedicating one's life to ending such atrocities, or serving God or humanity by helping others in the camps—all of which imply a plotline that can be imagined. The military has learned that soldiers show greater resilience even after having been horribly maimed in battle if they find a story that makes what happened to them meaningful in their eyes, so that they do not just feel horrible, victimized by fate.[3] Recent studies show that children whose families tell them stories about their history, with its ups and downs, have greater resilience, even when they experience significant trauma, such as a shooting in their school or a natural catastrophe.[4] Experts surmise that such family stories help children see difficulties as a normal part of life and have hope that they, as members of their families, will be able to cope. Similarly, we can increase our resilience by looking back over our lives to recognize the difficulties we have faced and the traumas we have endured, along with evidence of our strengths, courage, talents, creativity, and the simple ability to keep on trucking.

While stories can save or empower a life, they also can destroy it. A negative diagnosis by a doctor, based on statistical data that a person accepts as a prediction about their fate, might become a self-fulfilling prophecy, hastening death. Someone who loses a business or a spouse who repeatedly tells herself the story that she is a screwup and always will be may commit suicide, sink into depression, or just keep messing up. For this reason, it is important that we learn to reframe the internal narratives that spiral our thoughts downward, substituting ones that open up possibilities.

However, it is important not to overdo positivity. People whose stories about themselves are all rosy generally are in denial, and those who consistently brag about how great they are usually are out of touch with their ordinary humanity, and thus can collapse into shame when they make a mistake or move quickly into making excuses and blaming others for whatever occurred. The sign of a mature and resilient person is one whose inner and outer narratives naturally reflect both strengths and limitations and who are not afraid to fully feel moments of triumph, joy, humiliation, remorse, and so on. This accomplishment also helps us accept the positive qualities and imperfections in others, without having to label them as the good ones and bad ones, as in old TV scripts where the outlaws always wore black hats and the law-abiding cowboys white ones.

Meaning also comes to us from the call of the future, typically delivered through the vehicle of the imagination, as was true for the Holocaust survivors mentioned above. Most individuals who reflect on the meaning of their lives find that the answers come in our fantasies, which, if you notice your own, typically have narrative structures. We imagine what kind of person we want to be, how we want to act in the world, and what outcomes we hope to experience and why. Sometimes our stories include a clear sense of purpose—a vision of what we want to accomplish and what kind of person we want and need to be for this vision to be realized. Implicit in this is a plotline, just as in a movie or a novel, with a particular kind of character taking a specific kind of journey to reach a happy ending (their future vision). The more attentive we become to the quality of stories swirling around in our heads, the more likely it is that we can formulate narratives that result in the outcomes we desire.

In meditation, focusing on the breath and noticing stories without getting pulled into believing them is a great protector against accepting as true a story that can harm you. Even in the normal course of moving through your day, you can gain this benefit by remaining aware of the stories that are being replayed in your psyche. You also can be attentive to how these stories make you feel. A sign that your story is authentic is that it makes you feel a sense of rightness. In ordinary times, this can leave you calm and peaceful, in difficult times, feeling your authentic anger, sadness, or grief, and in good times, energized and excited.

So many men and women—and especially women—are being treated with antianxiety and antidepression drugs, it appears that we have an epidemic of quiet unhappiness on our hands in the United States today. If we add to the mix the many men and women who are self-medicating with alcohol and recreational drugs, those struggling with obesity from emotional eating and with other addictive behaviors, and the many people who attempt suicide, the numbers become staggering.

The prevailing medical story is that we need medication, which is correct with some cases of depression. However, too many of us take prescriptions or self-medicate rather than stopping to realize that we are unhappy about real things that we actually could change. It may be that all this stress and unhappiness means that the stories we are living, are being told, and tell ourselves are out of touch with reality or undermining our confidence. At the very least, we can stop believing and telling ourselves narratives that leave us insecure and breathless with anxiety and worry.

By clarifying who we are, our stories also help us know who we are not. Let's say you feel a moment of envy for another woman for qualities or accomplishments she has and you do not. While this can be a wake-up call that you want to develop those attributes in yourself, more frequently it provides an opportunity to notice how she shines, and then tell yourself that while these are not your gifts, you have equivalent ones, which you then can reinforce in your mind. This allows you to sort out your priorities, so that you no longer try to please everyone or struggle endlessly to live up to myriad idealized images of what women should be like that have little to do with who you are as an individual. When you know who you are, you more easily can have a happy and successful life without running yourself ragged.

As the previous chapters have illustrated, each archetype contributes to your finding a particular needed human dimension that can help you know who you are and what you are here to do: Demeter helps you know who and what you care about enough to care for. Zeus urges you to find your strengths and develop them so that you can contribute to the greater good as you also gain respect from others. Persephone connects you with deeper intuitive knowing of what you truly are called to do. And Dionysus seduces you into discovering what you enjoy doing for its own sake.

Your grail is the bliss you feel when you have gained mastery in all of these elements, so that you are showing up as your best and most complete and authentic self. The danger here is that archetypes often seem to want to be expressed fully in us, so mastery includes paring down your own expression of each archetype to what serves your uniqueness, priorities, and purpose.

Today, understanding these archetypes can transform what it means for women to "have it all," not just in the way men have done. Theoretically, men often have had jobs, families, and time for themselves, but that does not mean they all felt fulfilled in all of these roles. Quite a few men with very prestigious careers hardly knew their children, replaced their wives with younger ones when the opportunity presented itself, and in retirement had no intrinsic sense of their worth without their former titles and power. Many men had unfulfilling and often demeaning jobs. Some, of course, did and do have it all. They love their work, are friends with their coworkers, remain intimate with and committed to their spouses or partners, and are close to their children (if they have them) and other family members and friends. And what many women and men think would be having it all may differ widely from this stereotype about what "it all" is.

Multiple roles of various kinds require being in touch with archetypes that provide energy for their tasks. For many people today, the Zeus archetype provides the juice to compete at work, while Demeter energizes more caring responsibilities, Persephone fuels romantic and psycho-spiritual development, and Dionysus bestows that unexpected spurt of joy that often comes with a spontaneous, playful action. Without an archetype to support you in the major tasks you perform, they can feel like drudgery.

However, the good news is that you can awaken an archetypal story similar to the way you download a software application onto one of your computerized gadgets (laptop, smartphone, tablet, etc.), especially the ones that help you with your daily tasks. An archetypal download, however, is from your unconscious into your conscious mind. Reading about the archetypes in this book or elsewhere can trigger this process, especially if you begin to notice how you might be like the characters in question or have lived situations comparable to those they deal with. Just as you click on an icon to open a software application, you activate an archetype by starting to utilize its perspectives in how you think and what you do.

One woman reports that she consciously calls up her Zeus-warrior energy when going into a difficult work meeting, which is a way to remember that she needs to enter the room a bit armored and let the tough part of herself take the lead in how she looks, what she says, and how she reacts. Another woman, with little kids, shares how, when she gets home exhausted and wants to go to bed or just read a good book, she calls up the Demeter part of her, which she remembers is centered in her heart. She stops, breathes slowly in and out of her heart, recalls how much she loves her kids (thinking of them sleeping, looking innocent), relaxes her facial muscles through smiling, and greets her children with affection and love. Another, prone to feeling shy, hangs back for a bit when most of her friends are putting up the volleyball net, but then calls upon her Dionysian fun side to show up. She hangs out for a few minutes, imagining herself playing and being connected to her friends and how good that will feel, and when someone drops out, she happily steps into the game.

Becoming conscious of the archetypal stories available to you also can allow you to educate your archetypes about what you need from them to fulfill your roles and what expression best fits with your values, priorities, and energy level. To reinforce your understanding that the archetypes are parts of you, you can name them, and tell them (in your head, in a journal, or out loud) what you need or do not need them to do, imagining them nodding their heads in agreement (clearly thinking, "Got that"). This helps you integrate the archetypes into your personality and ego structure, so that their expressions all seem and are congruent with who you are. That way, you can stay energized in your many roles, without pushing yourself over the edge trying to be superwoman.

The Eleusinian Mysteries included all four archetypes because together they are more than their individual parts. The ultimate ideal is not just to have the energy for various tasks and roles but for all of your roles to be multidimensional expressions of who you are, so that you can show up as an integrated being everywhere. You do not have to check your caring, playfulness, or capacity for intimacy at the door when you go to work, and you do not have to stop thinking clearly, setting boundaries, and being assertive when you go home.

The Magic Wand That Shifts the Story

If we have the flexibility and creativity to imagine alternative stories, we can substitute any number of new narratives for tired, old, disempowering ones. When you change your inner scripts, you automatically change your behaviors, and then naturally change your life. Your magic wand is your capacity to transform the plotlines you are thinking, telling, and living, as well as the savvy that allows you to recognize what kind of script is needed for you to prevail in meeting the challenges of your life. This helps you become less rigidly locked into your own view of things and genuinely curious about how others see them. All of us have a partial view of the world, since we observe it through our own lenses, or habitual narratives, so we interpret external events to fit the plots we already have going on in our heads.

To get a more complete perspective, we need to expand what we are capable of seeing through listening to others. Think of how police officers responding to an accident or crime are supposed to determine what happened. Ideally, they gather together the often-conflicting testimony of eyewitnesses or others holding key information and only then make a judgment about what is true. So too with detectives and amateur sleuths who solve crimes by finding "who done it," or with us, when we are trying to decode the greater mystery of who we are and how we can make a difference.

Revolutions that forcibly overthrow the entrenched establishment typically result in narrowed, ideological thinking. Extreme examples are present in all totalitarian regimes that stifle potentially helpful counterviews through violent means. Even well-meant change efforts—whether with your spouse or partner, your larger family, your workplace, your place of worship, or a social unit—undertaken without genuine dialogue result in passive resistance, a general erosion of energy, or even the whole system being blindsided by unanticipated problems that some people understood would happen but feared to speak up about. Sometimes our personal "shoulds" become inner tyrants that repress our real feelings and our rebellious narratives.

The power of the magic wand starts with the cognitive, emotional, and narrative intelligence to formulate stories that bridge conflicting views.[5] Exploring events in your life through the lenses of the archetypal gods and goddesses of this book can facilitate this skill, especially if you also are aware of their narratives in the people around you. Doing this can remind you not to shoot them down if you disagree, but instead to listen for what you may be missing.

The next step in wise use of your magic wand requires claiming your power to tell it as you see it and cultivating the courage to enter a dialogue that may transform your views as well as those of others. And sharing your resulting narrative almost always will be more effective than arguing a position based on opinions or facts only, since stories provide context that explains why you think what you do. Leadership expert Stephen Denning (in *The Leader's Guide to Storytelling*) cites research demonstrating that we can provide people with all the data in the world on the need to change direction, even to the point that they may be able to pass a test on it, but then they likely will not act any differently than they did before. However, a powerful narrative that touches the heart can move people to reevaluate their situations and make changes.

Stories connect the heart and the mind in ways that spur action and often are remembered in memorable short phrases or sentences that evoke a larger narrative. Think of Lincoln's Gettysburg Address and his message that "It is . . . for us to be here dedicated to the great task remaining before us," and "we here highly resolve that these dead shall not have died in vain"; or Franklin Delano Roosevelt's First Inaugural Address, where he calmed a terrified nation, proclaiming that "the only thing we have to fear is fear itself."

Narrative-intelligence magic wands also can be part of everyday human exchanges. The chair of the graduate program in the English Department where I taught early in my career mentored me by giving me some advice. However, what sticks in my mind is not what I learned from what he told me, but what I learned through his example. I once watched anxiously as it looked like a major conflict was about to break out between him and an angry and challenging student. The chair paused and then calmly replied, "That is where you and I differ," and he went on, speaking softly while maintaining friendly eye contact to explain how he saw the situation. Just watching this expanded my sense of how to avoid escalating conflict. Many years later, I experienced another magic-wand epiphany while at lunch with a friend. I was complaining, in a poor-me tone of voice, about something my husband had said. Her response was, "Carol, it is just weather. When it rains, you don't cry; you get an umbrella." She shifted my story so quickly that I went from whimpering to laughing in an instant, remembering I actually was living a love story, where of course there are miscommunications that feel hurtful. If you read love stories, you see that such moments are as inevitable as fluctuations in the weather and can lead to the joys of making up.

The ability to tell a story from one's heart can have an especially powerful impact on the views of those around you, and if enough of you are telling a similar story, you collectively may be able to change the hearts and minds of your family, your boss, or even your nation. In the 1960s, a teenage girl who got pregnant would be in disgrace unless she managed to get an abortion, which was illegal and sometimes lethal. If she survived, she likely would keep it all secret, as would her family. If she continued the pregnancy, she had to drop out of school, possibly be destined to a life of poverty, and might have trouble marrying unless the father did the "honorable thing," in which case he might resent her and the child. Yet some very ordinary-seeming people pushed to change this. As a result, only the most hard-hearted person today would support such an outcome for young teens, however divided people might remain on issues of sex education or abortion. The more recent shift in public attitudes about gays and lesbians began when millions came out of the closet and talked openly about their inner and outer struggles, and movies such as *Brokeback Mountain* told stories that built further empathy. A similar movement is pending for our increasingly visible transgender population.

However inconsequential it feels when you just are sharing your truth, your narratives and modeling are leverage points from which you can begin to transform the world.

Martin Luther King Jr. provides a powerful example of narrative intelligence in practice. In his speeches and writings, especially to audiences that included whites as well as blacks, he first talked about personal experiences—for example, what it feels like as a parent when there is a playground and swimming pool in your neighborhood, but your kids are not allowed in—to evoke empathy, then moved to discuss the more general condition of African-Americans under segregation, and finally contextualized the civil rights movement in terms of people of all races being true to the founding principles of this nation. By building toward this larger story, he shifted the focus from a widely held cultural narrative of whites against blacks (still cherished by a few even today) to a call for all right-thinking Americans to unite in solidarity against ignorance and prejudice. President Obama (at the time of this writing) has continued in this tradition, sharing his story as he discusses the American dream as a vision still in the process of being realized as we form "a more perfect union."

Magic-wand stories are not just those we tell. Sometimes the most influential ones are those we live. A few years ago, at a time when I was running a leadership institute, experts in the field began to advocate the fine art of showing up in the fully embodied presence of your authenticity. What is now known as "presencing" is as relevant to anyone, whatever our social roles, as it is to leaders. The book that put this on the leadership agenda, *Presence: Human Purpose and the Field of the Future,* was written by a high-powered organizational development team that included Peter Senge, C. Otto Scharmer, Joseph Jaworski, and Betty Sue Flowers, who share with the reader key moments in their own journeys, thus modeling transparency in exploring the deeper parts of leadership, in contrast to the widespread focus in career development books on money, power, and status.

As accomplished as all four authors are, they take for granted that leaders need to demonstrate strength, competence, and care for others. *Presence* maintains that individuals have a purpose, so that their living authentically is important to the whole, and each of us is a microcosm of that whole (similar to Hushpuppy's metaphysics). To connect with the world in this way requires that your authentic and deeper self be embodied in your physical presence, your actions, and everything you do. It also allows you to be spontaneous, because you are not hiding who you are, and enjoy your relationship with others. In this, we can hear the caring of Demeter, the power of Zeus, the deep authenticity of Persephone, and the freedom of Dionysus. It means that you can be fully you as a leader and have the flexibility to live any one of these stories as required by circumstances. Such authentic situational leadership allows you to transform events and relationships.

Sometimes this shift can be achieved just through body language. For example, if someone comes into your office angry, welcoming them with a Demeter-caring affect, with arms outstretched and a wide, authentic smile, often can transform their mood. However, if they go over the line in what they say or do, changing your posture to be more defended, looking strong with some Zeus-like element of warning in your eyes, may get them to show you more respect. A bit of Dionysian playfulness in how you are acting can lighten things up, while Persephone's dignified stillness can calm them down. In a more macro way, the art of presencing comes from modeling the story you desire to evoke in others or in the larger social system. Whatever story you tell, people need to see you live it to believe in you and be willing to join you in it.

Each of the archetypal stories featured in the preceding chapters has contained narrative intelligence lessons. All of the characters utilize their inner swords to free themselves from limiting stories, their shields to protect them from ongoing pressure from others, their grails in solidifying their actions around their true archetypal purposes, and in all this, their magic wands to shift the larger story around them. That is how community is restored and their happy endings are achieved.

A final narrative-intelligence ability allows us to get to the point and explain the essence of the narratives we are living and telling. Because some people get lost in the details, many teaching stories classically end with "and the moral of this story is _____." I know that I appreciate it when an author crystallizes a book's core message at its end. So here goes: The metanarrative that emerges from *Persephone Rising* tells you that you matter—we all do—and that when you are the most true to yourself and your calling, you do the most good. For millennia, humans have been sold a narrative that tells us that some of us are the main characters in the story and others are but supporting characters, meant to serve. Along with this, the idea has been perpetuated that what women care about, do, and have done is of lesser value than male priorities and behaviors. But now we are in a new situation, where we cannot afford the loss of human capacities that such stories have fostered—about gender, race, income, sexual orientation, or any of the factors that have been used to marginalize and devalue some of us for the benefit of others. As individuals, groups, and whole societies we can refuse to buy into and agree with the stories that disempower us. We can recognize that they are only stories, and their power comes from people acting as if they were true, so we should stop reinforcing what we do not want.

Throughout this book, you have discovered an alternative ancient narrative—the Eleusinian Mysteries—and seen how the archetypes in them are alive today, reemerging in behaviors that often are regarded as unrelated to one another. These archetypal narratives of the gods and goddesses of this Eleusinian Mystery tradition also have demonstrated how you can connect with key inner resources so that you are supported by an awakened curious and questioning mind, a brave and caring heart, the wise guidance of your soul, and the joy that can be found in a body that feels alive and free. All of these are your birthright, but you must claim them. This is the heroic act that allows you to transform your story and your life.

As you are more and more supported by these inner capacities, a side product is that you can remain centered in yourself while also feeling connected with others—from those closest to you emanating out in widening spirals to include the human species, the earth, the cosmos. If enough of us do this, our combined human capacity will allow us to transform our world, freeing us from the cynicism and despair that currently is limiting human potential.

Stories throughout this book also have revealed how archetypes in individuals and in the larger society mirror one another in an interdependent way that allows for individual, group, and human evolution. If you take this in and think about it, it can reassure you that you are part of complex, interconnected social and natural living systems, the center of which is nowhere and everywhere. For you, it is wherever you are. Anytime you change, you cause ripple effects that disturb the systems around you that then readjust, creating a new order, which in turn influences you, and so on and so on. Thus, your seeking authentic fulfillment contributes to a complex process that helps larger social and environmental systems evolve along with you.

Whatever your metanarrative, my hope for you is that you find joy, experience abundance, and demonstrate fearlessness in your fidelity to your own life path—gaining the rewards promised to ancient initiates over two millennia ago and even more fully attainable today.

Who's Who

THE RELATIONSHIPS AMONG the gods and mortals in Greek mythology are often confusing and frequently inconsistent, depending on the source. This Who's Who is designed to help you keep track of the figures discussed in this book. The Roman equivalents are provided where relevant.

Aphrodite—Goddess of love, beauty, pleasure, and procreation; wife of Hephaestus; known as Venus in Roman mythology.

Apollo—God of the sun, music, poetry, and many other things; son of Zeus; patron of Delphi.

Ariadne—Wife of Dionysus; daughter of King Minos and Queen Pasiphae of Crete; half-sister of the Minotaur; helped Theseus kill the Minotaur.

Athena—Goddess of wisdom and vehicles; daughter of Metis by Zeus; protector of Athens.

Cronus—King of the Titans; husband of Rhea; father of Zeus, Hera, Hades, Poseidon, Demeter, and Hestia.

Demeter—Mother of Persephone by Zeus; daughter of the Titans Cronus and Rhea; sister of Zeus; founder of the Eleusinian Mysteries.

Demophon—Son of Queen Metaneira.

Dionysus—God of the grape harvest, winemaking and wine, ritual madness, fertility, theater, and religious ecstasy; son of Persephone and Hades; husband of Ariadne; often conflated with Bacchus, the Roman god of wine.

Eros—God of love; husband of Psyche; known as Cupid in Roman mythology.

Gaia—The primal Earth Mother.

Hades—God of the Underworld; husband of Persephone; son of the Titans Cronus and Rhea; brother of Zeus, Demeter, Poseidon, Hera, and Hestia.

Hekate—Goddess of the crossroads; shows kindness to Demeter; takes Persephone's place in the Underworld when she is absent.

Hephaestus—God of blacksmiths, craftsmen, artisans, and other things; son of Hera by parthogenesis; husband of Aphrodite.

Hera—Wife of Zeus; mother of Hephaestus by parthogenesis; daughter of the Titans Cronus and Rhea; sister of Zeus, Poseidon, Demeter, Hades, and Hestia; frequently portrayed as weeping, jealous, or angry.

Hermes—God of commerce and communication; accompanies Persephone back from the Underworld.

Iambe—Servant of Queen Metaneira; also the goddess of humor and poetry.

Kore/Persephone—Queen of the Underworld; daughter of Demeter and Zeus; wife of Hades; founder of the Eleusinian Mysteries.

Metaneira—Queen of Eleusis.

Metis—Titan goddess; advisor to Zeus; mother of Athena.

Minotaur—Offspring of Queen Pasiphae of Crete and a bull.

Poseidon—God of the seas; son of the Titans Cronus and Rhea; brother of Zeus, Hades, Demeter, Hera, and Hestia.

Psyche—Wife of Eros; mother of Hedone.

Rhea—Wife of Cronus; mother of Zeus, Demeter, Poseidon, Hades, Hera, and Hestia.

Typhon—Monster created by the Titans to defeat Zeus.

Zeus—Chief god of Olympus; god of the sky; son of the Titans Cronus and Rhea; husband of Hera; father of Persephone by Demeter (and many other illegitimate children); brother of Hera, Demeter, Poseidon, Hades, and Hestia.

Bibliography

Essential Sourcebooks About the Eleusinian Mysteries and Related Myths

Blackford, Holly Virginia. *The Myth of Persephone in Girls' Fantasy Literature*. New York: Routledge, 2012.

Bowden, Hugh. *Mystery Cults of the Ancient World*. Princeton, NJ: Princeton Univ. Press, 2010.

Demand, Nancy. *Birth, Death, and Motherhood in Classical Greece*. Baltimore: Johns Hopkins Univ. Press, 2004.

Downing, Christine, ed. *The Long Journey Home: Re-Visioning the Myth of Demeter and Persephone for Our Time*. Boston: Shambhala Publications, 1994. If you want to follow up with only one book, I'd recommend this one.

——. *Psyche's Sisters: Reimagining the Meaning of Sisterhood*. Reprint, New Orleans: Spring Journal Books, 2007.

Ehrenreich, Barbara. *Dancing in the Streets: A History of Collective Joy*. New York: Metropolitan Books, 2007.

Eisler, Riane. *The Chalice and the Blade: Our History, Our Future*. Cambridge, MA: Harper & Row Publishers, 1987.

Foley, Helene P., ed. *The Homeric Hymn to Demeter: Translation, Commentary, and Interpretive Essays*. Princeton, NJ: Princeton Univ. Press, 1994.

Frazer, James George. *The Golden Bough: A Study in Magic and Religion*. New York: Macmillan Publishing, 1922.

Gadon, Elinor W. *The Once and Future Goddess*. New York: Harper & Row Publishers, 1989.

Gilligan, Carol. *The Birth of Pleasure*. New York: Knopf, 2002.

Hillman, James. *The Myth of Analysis: Three Essays in Archetypal Psychology*. Evanston, IL: Northwestern Univ. Press, 1983.

Johnson, Robert A. *Ecstasy: Understanding the Psychology of Joy*. San Francisco: Harper & Row Publishers, 1987.

Keller, Mara Lynn. "The Ritual Path of Initiation into the Eleusinian Mysteries." *Rosicrucian Digest* 2 (2009): 28–42.

Kerényi, Carl. *Dionysos: Archetypal Image of Indestructible Life*. Translated by Ralph Manheim. Vol. 2 of *Archetypal Images in Greek Religion*. Mythos 65. Series sponsored by Bollingen Foundation. Princeton, NJ: Princeton Univ. Press, 1976.

Kingsley, Peter. *In the Dark Places of Wisdom*. Inverness, CA: The Golden Sufi Center, 1999.

Louis, Margot K. *Persephone Rises, 1860–1927: Mythography, Gender, and the Creation of a New Spirituality*. Farnham, UK: Ashgate, 2009.

Swanson, Todd. "Womb of Fire: A Study of the Eleusinian Mysteries," *Eleusinian Mysteries* (blog). May 1993. http://eleusinianmysteries.org/WombOfFire.html.

Wilkinson, Tanya. *Persephone Returns: Victims, Heroes, and the Journey from the Underworld*. Berkeley, CA: PageMill Press, 1996.

Additional Highly Recommended Books

Adson, Patricia R., and Jennifer Van Homer. *A Princess and Her Garden: A Fable of Awakening and Arrival*. 2nd ed. Gainesville, FL: Center for Applications of Psychological Type, 2011.

Arbinger Institute, The. *Leadership and Self-Deception: Getting Out of the Box*. San Francisco: Berrett-Koehler Publishers, 2009.

Bastian, Edward W. *InterSpiritual Meditation: A Seven-Step Process from the World's Spiritual Traditions*. Santa Barbara, CA: Spiritual Paths Publishing, 2010.

———. *Mandala: Creating an Authentic Spiritual Path: An InterSpiritual Process*. Boulder, CO: Albion, 2014. Includes an instrument to find your natural spiritual path.

Beckwith, Michael Bernard. *TranscenDance*. Culver City, CA: Agape Media International (distributed by Hay House), 2012.

Bolen, Jean Shinoda. *Artemis: The Indomitable Spirit in Everywoman*. San Francisco: Conari Press, 2014.

———. *Goddesses in Everywoman: A New Psychology of Women*. San Francisco: Harper & Row Publishers, 1984.

Campbell, Joseph, and Bill Moyers. *The Power of Myth*. New York: Doubleday, 1988.

Casey, Caroline W. *Making the Gods Work for You: The Astrological Language of the Psyche*. New York: Harmony Books, 1998.

Colman, Arthur D. *Up from Scapegoating: Awakening Consciousness in Groups*. Wilmette, IL: Chiron Publications, 1995.

Edinger, Edward F. *Ego and Archetype: Individuation and the Religious Function of the Psyche*. Boston: The C. G. Jung Foundation for Analytical Psychology, 1972.

Estés, Clarissa Pinkola. *Women Who Run with the Wolves: Myths and Stories of the Wild Woman Archetype*. New York: Ballantine Books, 1992.

Fox, Matthew. *The Hidden Spirituality of Men: Ten Metaphors to Awaken the Sacred Masculine*. Novato, CA: New World Library, 2008.

———. *Original Blessing: A Primer in Creation Spirituality Presented in Four Paths, Twenty-Six Themes, and Two Questions*. Santa Fe, NM: Bear, 1983.

Gimbutas, Marija. *The Language of the Goddess*. San Francisco: Harper & Row Publishers, 1989.

Hale, Cynthia Anne. *The Red Place: Transforming Past Traumas Through Relationships*. London: Muswell Hill Press, 2014.

Heilbrun, Carolyn G. *Toward a Recognition of Androgyny: A Search into Myth and Literature to Trace Manifestations of Androgyny and to Assess Their Implications for Today*. New York: Alfred A. Knopf, 1973.

Henderson, Hazel. *Paradigms in Progress: Life Beyond Economics*. Indianapolis: Knowledge Systems, 1991.

Hillman, James. *Re-Visioning Psychology*. New York: Harper & Row Publishers, 1975.

Houston, Jean. *The Search for the Beloved: Journeys in Mythology and Sacred Psychology*. Los Angeles: Tarcher/Perigee, 1987.

Huffington, Arianna. *Thrive: The Third Metric to Redefining Success and Creating a Life of Well-Being, Wisdom, and Wonder*. New York: Harmony Books, 2014.

Johnson, Robert A. *Inner Work: Using Dreams and Active Imagination for Personal Growth*. San Francisco: Harper & Row Publishers, 1986.

Jung, C. G., Gerhard Adler, and R. F. C. Hull (eds.), *Archetypes and the Collective Unconscious*, from the Collected Works of C. G. Jung, Volume 9, Part 1. Princeton, NJ: Princeton Univ. Press, 1981.

Kay, Katty, and Claire Shipman. *The Confidence Code: The Science and Art of Self-Assurance—What Women Should Know*. New York: HarperBusiness, 2014.

King, Vivian. *Soul Play: Turning Your Daily Dramas into Divine Comedies*. Georgetown, MA: Ant Hill Press, 1998.

Korten, David. *Change the Story, Change the Future: A Living Economy for a Living Earth*. San Francisco: Berrett-Koehler Publishers, 2015.

L'Engle, Madeleine. *A Wind in the Door*. New York: Farrar, Straus and Giroux, 1973.

Louden, Jennifer. *The Life Organizer: A Woman's Guide to a Mindful Year*. Novato, CA: New World Library, 2007.

Lule, Jack. *Daily News, Eternal Stories: The Mythological Role of Journalism*. New York: Guilford Press, 2001.

MacCoun, Catherine. *On Becoming an Alchemist: A Guide for the Modern Magician*. Boston: Trumpeter, 2008.

Mahaffey, Patrick. *Evolving God-Images: Essays on Religion, Individuation, and Postmodern Spirituality*. Bloomington, IN: iUniverse, 2014.

Moore, Robert L., and Douglas Gillette. *King, Warrior, Magician, Lover: Rediscovering the Archetypes of the Mature Masculine*. San Francisco: HarperSanFrancisco, 1991.

Murdock, Maureen. *The Heroine's Journey: Woman's Quest for Wholeness*. Boston: Shambhala Publications, 1990.

Noble, Vicki. *Motherpeace: A Way to the Goddess Through Myth, Art, and Tarot*. San Francisco: Harper & Row Publishers, 1983.

Paris, Ginette. *Heartbreak: New Approaches to Healing; Recovering from Lost Love and Mourning*. Minneapolis: World Books Collective, 2014.

———. *Pagan Grace: Dionysos, Hermes, and Goddess Memory in Daily Life*. Dallas, TX: Spring Publications, 1991.

Rosas, Debbie, and Carlos Rosas. *The Nia Technique: The High-Powered Energizing Workout That Gives You a New Body and a New Life.* New York: Broadway Books, 2004.

Roth, Gabrielle. *Sweat Your Prayers: Movement as Spiritual Practice.* New York: Jeremy P. Tarcher/Putnam, 1998. Of her five rhythms: flowing is the rhythm of Demeter, staccato of Zeus, chaos of Dionysus, and stillness of integration of these in your core.

Roth, Gabrielle, with John Loudon. *Maps to Ecstasy: The Healing Power of Movement.* 2nd ed. Novato, CA: New World Library, 1998.

Sahtouris, Elisabet. *Gaia: The Human Journey from Chaos to Cosmos.* New York: Pocket Books, 1989.

Sandberg, Sheryl. *Lean In: Women, Work, and the Will to Lead.* New York: Alfred A. Knopf, 2013.

Starhawk. *Dreaming the Dark: Magic, Sex and Politics.* Boston: Beacon Press, 1982.

Stone, Hal, and Sidra Stone. *Embracing Our Selves: The Voice Dialogue Manual.* Novato, CA: New World Library, 1989.

Tarnas, Richard. *The Passion of the Western Mind: Understanding the Ideas That Have Shaped Our World View.* New York: Ballantine Books, 1991.

Von Franz, Marie-Louise. *Alchemy: An Introduction to the Symbolism and the Psychology.* Toronto: Inner City Books, 1980.

Woodman, Marion, and Elinor Dickson. *Dancing in the Flames: The Dark Goddess in the Transformation of Consciousness.* Boston: Shambhala Publications, 1996.

Woolger, Jennifer Barker, and Roger J. Woolger. *The Goddess Within: A Guide to the Eternal Myths That Shape Women's Lives.* New York: Fawcett Columbine, 1989.

Related Publications by Carol S. Pearson

Awakening the Heroes Within: Twelve Archetypes to Help Us Find Ourselves and Transform Our World. San Francisco: HarperSanFrancisco, 1991. Translations: Chinese, German, Italian, Japanese, Portuguese, and Spanish.

Educating the Majority: Women Challenge Tradition in Higher Education, coeditors, Donna L. Shavlik and Judith G. Touchton. New York: Macmillan Publishing, 1989.

The Female Hero in American and British Literature, coauthor, Katherine Pope. New York: R. R. Bowker, 1981.

The Hero Within: Six Archetypes We Live By. San Francisco: HarperSanFrancisco, 1986, rev. eds. 1989, 1998. Translations: Danish, Dutch, French, German, Italian, Korean, Polish, Portuguese, Spanish, and Turkish.

Introduction to Archetypes, coauthor, Hugh Marr. Gainesville, FL: Center for Applications of Psychological Type, 2002.

Magic at Work: Camelot, Creative Leadership, and Everyday Miracles, coauthor, Sharon Seivert. New York: Doubleday/Currency, 1995. An earlier book that explores the Arthurian stories of Camelot for their applications in the modern world.

Mapping the Organizational Psyche: A Jungian Theory of Organizational Dynamics and Change, coauthor, John G. Corlett. Gainesville, FL: Center for Applications of Psychological Type, 2003.

The Pearson-Marr Archetype Indicator™, coauthor, Hugh Marr. Gainesville, FL: Center for Applications of Psychological Type, 2003, new edition 2014. This instrument provides feedback on your heroic archetypes, which helps you to know your underlying values, motivation, and narrative mind-set. Unlike many surveys or questionnaires, this is a well-tested instrument, designed by an expert psychometric and theoretical team. In concert with *Persephone Rising*, it can show you where you are in the journeys of Zeus from Orphan to Warrior to Ruler; of Demeter from Caregiver to Seeker to Sage; of Persephone from Innocent to Lover to Magician, and of Dionysus from Destroyer to Jester to Creator (archetypal names found in *Awakening the Heroes Within*). To take the instrument, go to www.capt.org, where you can find additional supporting materials and books.

The Transforming Leader: New Approaches to Leadership for the Twenty-First Century, ed. San Francisco: Berrett-Koehler Publishers, 2012.

Who Am I This Time? Female Portraits in British and American Literature, coauthor, Katherine Pope. New York: McGraw-Hill, 1976.

Carol S. Pearson's Archetypal Theories Applied to Psychology, Coaching, and Branding

Adson, Patricia R. *Depth Coaching: Discovering Archetypes for Empowerment, Growth, and Balance.* Gainesville, FL: Center for Applications of Psychological Type, 2004.

——. *Finding Your Own True North and Helping Others Find Direction in Life.* Gladwyne, PA: Type & Temperament, 1999.

Atlee, Cindy. *Becoming Known Well: Authentic Personal Branding for Workplace Fulfillment, Success and Contribution.* n.p.: Storybranding Group, 2013.

——. *Using Narrative Intelligence to Lead, Motivate and Communicate.* n.p.: Storybranding Group, 2014.

——. *When the Product is You: Branding from Authentic Self.* n.p.: Storybranding Group, 2012.

Mark, Margaret, and Carol S. Pearson. *The Hero and the Outlaw: Building Extraordinary Brands Through the Power of Archetypes.* New York: McGraw-Hill, 2001. Paperback edition released September 2002. Translations: Chinese, Estonian, Portuguese, and Russian.

For additional books, resources, services, blogs, and an opportunity to engage with the author and her colleagues, go to www.herowithin.com. You also can follow her on LinkedIn, Facebook: *Carol S. Pearson, PhD,* and Twitter.

About the Author

Carol S. Pearson, Ph.D., D.Min., is an internationally known authority on archetypes and their application to everyday life and work. Through increasing narrative intelligence, her work has helped people all over the world to find their purpose and live more fulfilling and successful lives, and organizations to realize their higher purpose. Best known for groundbreaking publications such as *The Hero Within: Six Archetypes We Live By; Awakening the Heroes Within: Twelve Archetypes to Help Us Find Ourselves and Transform Our World; The Transforming Leader: New Approaches to Leadership for the Twenty-First Century* (ed.); *The Hero and the Outlaw: Building Extraordinary Brands Through the Power of Archetypes* (coauthor, Margaret Mark); and *The Pearson-Marr Archetype Indicator*™ (coauthor, Hugh Marr), her writing is informed and tested by both her personal and high-level professional experience.

A respected scholar and higher education administrator, Dr. Pearson served most recently as executive vice president / provost and then president of Pacifica Graduate Institute. Previously, she was professor of leadership studies in the School of Public Policy at the University of Maryland and director of the James MacGregor Burns Academy of Leadership. During her tenure, the academy was the incubator of the

International Leadership Association (ILA), and Dr. Pearson was a member of ILA's board of directors. Earlier in her career, she served as the director of two major women's studies programs and the academic dean of a women's college. In collaboration with the Office on Women at the American Council on Education, she helped construct guidelines for how colleges and universities could better serve the needs of women administrators, faculty, and students.

Currently, Dr. Pearson is an author, speaker, workshop leader, and consultant in private practice. In the latter role, she serves as a thinking partner to leaders and leadership teams in the not-for-profit, educational, political, and governmental sectors and mission-driven, triple-bottom-line companies on areas such as building and managing heroic enterprises, archetypal reputation management, and message and leadership development. As a workshop leader, she offers a variety of leadership development experiences as well as intensives on awakening the hero and heroine within.

Website: www.herowithin.com.

You also can follow Dr. Pearson on LinkedIn (www.linkedin.com /pub/carol-pearson/21/81a/a4a); Facebook: *Carol S. Pearson, PhD;* and Twitter (@carolspearson).

Acknowledgments

The pioneering psychiatrist C. G. Jung defined synchronicity as a "meaningful coincidence"—two or more events that appear to be meaningfully but not causally related.

Early in 2013, when I was in the process of leaving my position as president of Pacifica Graduate Institute and journaling about my next steps, I found myself writing instead about the story of Demeter and Persephone. Then, seemingly out of the blue, I received an e-mail from Claudia Boutote, senior vice president and associate publisher of HarperOne (formerly HarperSanFrancisco), the publisher of my earlier books *The Hero Within* and *Awakening the Heroes Within*, asking me to "return home" and write for HarperOne once again.

The timing was so synchronistic that I soon arranged to visit the HarperOne offices in San Francisco. When I walked in, I saw *The Hero Within* prominently displayed in the reception area—which I suspect is HarperOne's usual practice when welcoming one of its authors. Claudia and I first met with HarperOne publisher Mark Tauber and then had a lovely lunch with senior vice president and executive editor Michael Maudlin. We quickly achieved a meeting of the minds about my ideas, augmented by theirs, for this book. From these conversations, it also became apparent that HarperOne remains a publisher that truly cares about books and the authors who create them, and one that promotes a mutually supportive team atmosphere that ameliorates the essential loneliness of the writing process.

While drafting the book proposal, I was delighted to discover that my former literary agent, Stephanie Tade, had founded her own company (the Stephanie Tade Agency) and was happy to represent me again. As a result of her expert guidance, within a few months, I secured a contract that was satisfactory to all parties. Then, of course, came the research and writing.

Like many narratives, however, this one actually begins long before I was contacted by HarperOne. As described in the book, the subject of *Persephone Rising* called me when I was in my early thirties and dealing with a major loss. But to write the book, I needed not just to be invested in the topic but also to be well trained in narrative intelligence and archetypal psychology. Thus, the first debt of gratitude I owe is to C. G. Jung for his revolutionary work on archetypes and myth, and to those who followed in his wake—Joseph Campbell, James Hillman, Marion Woodman, and many others—some of whom I have had the privilege to know personally.

Synchronicity continued to work for me in my training to write this book. I was introduced to Jung, Campbell, archetypes, and myth at Rice University, where I received my undergraduate and graduate degrees during the relatively short window of time when the myth and symbol school of literary criticism was emphasized there. I did most of my Doctorate of Ministry at the University of Creation Spirituality when it emphasized ecumenical creation spirituality, and finished after it was renamed Wisdom University, studying wisdom literature and its applications. Together, these curricula were essential in preparing me for this project. The Midway Center for Creative Imagination, which provided training in the uses of active imagination; the Center for Advanced Studies in Depth Psychology Professional Enrichment Program in Jungian Theory and Practice, where I received postgraduate training in depth psychology; and Nia dance-teacher trainings, which helped me access the archetypes in my body—all were available just when I would benefit from them most.

As a college and university faculty member and administrator, it seemed I always was where I could learn the most at that time, often from my colleagues and students. I was fortunate to have been the first director of women's studies programs at the University of Colorado and later at the University of Maryland, where we incubated the National Women's Studies Association. My wonderfully radical colleagues at the University of Colorado also helped me understand the importance of Dionysian wildness. Did we ever have fun! More recently, I was the beneficiary of wisdom that rubbed off on me from the faculties of the School of Public Policy at the University of Maryland and of Pacifica Graduate Institute.

I could not have written this book without the work of many distinguished scholars, several of whom were colleagues of mine at Pacifica, who have done research on the Eleusinian Mysteries. I am grateful I had the benefit of access to the abundant resources of the Pacifica Research Library mythology and depth psychology collections, which facilitated access to this rich literature. I also benefitted from the competence of the library staff and how they provided ease in utilizing the collection.

Writing a book is lonely work, solitary by nature, and a subject such as this one tends to pull one inward, at least it did for me. However, I am grateful for how much support I had from others. Five people read my initial draft and helped to improve it, especially Cynthia Hale, who gave detailed feedback on an earlier version of the manuscript. Pat Adson and Cindy Atlee contributed valuable overall input, and JoAn Herren and Ed Bastian shared insights on their own "Zeus within" when I was revising his chapter. Others who provided me with just the insight I needed at a moment in time include Thomas Wilkenson, who pointed me to the Seattle Seahawks example in the Dionysus chapter; James Palmer, who offered insights into *The King's Speech;* Aryeh Maidenbaum and Diana Rubin, directors of the New York Center for Jungian Studies, where I gained important insights on the relationships between fate, choice, and destiny; and Michael Conforti, director of the Assisi Institute, whose seminars enhanced my understanding of his work on archetypal patterns and processes, especially through his unflinching recognition of their pathological sides.

During the writing process, I also received assistance from Megan Scribner, who, at two critical junctures, helped me improve continuity and after I had seriously overwritten, cut the manuscript down to a manageable size, as she had done with my previous book, *The Transforming Leader;* Mark Kelly, who helped to find references for insights and facts that I remembered but for which I had no detailed record; and Dorene Koehler, my virtual assistant, who resolved multiple issues with endnotes, including figuring out what to do when my Word software jumbled their numbering. Michael Maudlin of HarperOne offered numerous observations and suggestions that sharpened the focus of the book.

As an additional fortuitous happening, just as *Persephone Rising* was nearing completion, Claudia Boutote announced the launch of an exciting new line, called HarperElixir, with a mind-body-spirit theme, and I was honored to learn that my book would be one of its initial offerings. At this point, I was assisted by Libby Edelson, senior editor of this new division, who oversaw the final structure and flow of the book and who also demonstrated a sharp eye for where I was being overly vague or making intuitive leaps that would leave the reader wondering, "What happened?" Thanks also to Lisa Zuniga, production editor; copyeditor Diane Huskinson; proofreader Tanya Fox; and the fine Harper marketing team.

Throughout the year and a half I spent researching and writing *Persephone Rising,* my incredible extended family and friendship network, as well as the local Nia dance teachers and community in Santa Barbara, kept me happy, sane, and grounded. My husband, David Merkowitz, did fact checking and major close editing, as he always does with my publications, while providing moral support as unresolved life issues surfaced, triggered by the content of the book. We have adjoining offices, so when I'm writing and he is editing, it is very companionable—except for his occasional groans of despair when my writing is not up to his standards. When HarperOne asked me to provide a manuscript by the second week in January to meet the

HarperElixir launch deadline, requiring us to cancel holiday plans, he didn't even complain, nor did my daughter, to whom this book is dedicated and who had been planning to visit us with her family.

David and I are incredibly lucky that all our children are good, caring people, doing work that makes a difference in the world, having married equally wonderful spouses. We love observing how each of these three couples runs their lives as equal partners and what great parents they are. And, of course, we could not be happier to now have six amazing grandchildren. I have learned a great deal about the subject of this book by seeing how our children live and by learning from their various perspectives on life, work, and love. And finally, I want to acknowledge the role of my parents in giving me a good start in life, and my brother, John Douglas, who consistently helps me lighten up, stay positive, and keep a sense of humor.

To all of the above, and to the many others who have provided moral, spiritual, and practical support and guidance, I offer my profound gratitude.

That's not the end of the story, however. As I send *Persephone Rising* out into the world, I am aware that it is you, the readers, who have been constantly in my thoughts and have inspired me to share whatever knowledge and insights I possess. Over the years, many readers of my books have written or e-mailed me, or approached me personally, to tell their stories and reveal how their lives have changed as a result of what they learned. Whether you are new to my work or a returning reader, I look forward to further such conversations, and I welcome whatever reactions you care to offer.

Notes

The Eleusinian Mysteries and the Power of Collective Transformation

1. Some sources believe that the sacrifice of the pigs happened at the earlier February rites. More generally, there are debates about exactly what happened where, since our ideas about the rites have been pieced together by scholars from divergent and often fragmentary sources.
2. Hugh Bowden, *Mystery Cults of the Ancient World* (Princeton, NJ: Princeton Univ. Press, 2010), 26.
3. Elinor W. Gadon, *The Once and Future Goddess: A Symbol for Our Time* (New York: Harper & Row Publishers, 1989), 143, 144, quotations. Her discussion of the rite, what happened there, and its connection with Minoan Crete can be found on pages 143–66.
4. Gadon, *Once and Future*, 26.
5. Helene P. Foley, ed., "Background: The Eleusinian Mysteries and Women's Rites for Demeter," *The Homeric Hymn to Demeter: Translation, Commentary, and Interpretive Essays* (Princeton, NJ: Princeton Univ. Press, 1994), 65–75.

Why Me? How This Tradition Changed My Life

1. John R. Haule, *Jung in the 21st Century. Vol. 1: Evolution and Archetype* (New York: Routledge, 2011), 6. According to Haule, "Jung believed that archetypes were embedded in evolutionary development." However, Haule continues, "not Jung, but Robin Fox, an anthropologist at Rutgers University, said: 'What we are equipped with is innate propensities that require environmental input for their realization' (Fox, 1989: 45). Fox insists that no account of the human condition can be taken seriously if it ignores the five million years of natural selection that have made us what we are (Ibid., 207). He lists more than twenty human patterns that would be sure to manifest if some new

Adam and Eve were allowed to propagate in a universe parallel to ours. These would be archetypal realities, passed on through DNA, and expressed in distinctive neuronal tracts in their brains. Such behavioral patterns would surely include customs and laws regarding property, incest, marriage, kinship, and social status; myths and legends; beliefs about the supernatural; gambling, adultery, homicide, schizophrenia, and the therapies to deal with them."

Why You? Awakening Your Capacities and Potential

1. Elgin's use of "intelligent design" refers to scientific findings that there appears to be an intelligence in nature. The phrase now has associations with particular political and denominational views not implied in Elgin's work. What he sees are recurrent physical structures in nature that reinforce psychological structures in people, which here are called archetypes.
2. Duane Elgin, *Awakening Earth: Exploring the Evolution of Human Culture and Consciousness* (New York: William Morrow, 1993), 25.

Why Now? Thriving in an Unfinished Revolution

1. The term *myth* also is used to mean "falsehood" and was employed to denigrate the wisdom of traditions seen as more primitive cultures than one's own, although, when most Greeks first converted to Christianity, many were initiates, too, seeing no conflict between the two sets of teachings. However, over time, Christian clerics began to regard other sources of wisdom as undermining the truth of what they were teaching and thus denigrated them as falsehoods.

Demeter and the Way of the Heart

1. N. F. Cantor, ed., *The Jewish Experience* (New York: HarperCollins, 1996), 124–28. In his essay on "The Essentials of Hasidism," Martin Buber mentions the Indian concept of "tat twam asi" [Thou art that]: "A saying dating back to the Baalshem himself as again related to the commandment to love thy neighbor" as thyself: [quoting Baal Shem Tov] "For every man in Israel has a root in the Unity, and therefore we may not reject him 'with both hands,' for whoever rejects his companion rejects himself: to reject the minutest particle of the Unity is to reject it all" (125).
2. Kabir Helminski, "Sayings of Muhammad: Selected and Translated by Kabir Helminski," *The Knowing Heart: A Sufi Path of Transformation* (Boston: Shambhala, 1999), 178–80.
3. Jalâl al-Dîn Rûmî and M. Green, *One Song: A New Illuminated Rumi,* trans. C. Barks (Philadelphia: Running Press, 2005), 11.
4. T. Piyadassi, trans., "Discourse on Loving-Kindness (Karaniya Metta Sutta)," *The Book of Protection* (Paritta) (Kandy, Sri Lanka: Buddhist Publication Society, 1975), 35–36.
5. Swami Sivananda, *Sivananda News* #11 (2003), www.sivananda.org/publications /mailinglists/guru-gram/gg-Nov01-2003.html. See also Yudit Kornberg Greenberg, ed., *Encyclopedia of Love in World Religions, Vol. 1* (Santa Barbara, CA: ABC-CLIO, 2008) for more information on the injunction to love thy neighbor, charity, divine love, festivals of love, and romantic love in all the world's major religions.
6. Mara Lynn Keller, "The Ritual Path of Initiation into the Eleusinian Mysteries," *Rosicrucian Digest* 2 (2009), 1.

7. Keller, "Ritual Path," 38.

8. See HeartMath, "HeartMath Science and Research," accessed May 17, 2015, www
.heartmath.com/research for research findings and https://www.heartmath.org for
more information on the HeartMath Institute.

9. Jonathan Reams, "Integral Leadership: Opening Space by Leading Through the
Heart," *The Transforming Leader: New Approaches to Leadership for the Twenty-First Century*,
Carol S. Pearson, ed. (San Francisco: Berrett-Koehler Publishers, 2012), 106–7.

10. Karin Grossmann, et al., "A Wider View of Attachment and Exploration: The Influ-
ence of Mothers and Fathers on the Development of Psychological Security from
Infancy to Young Adulthood," *Handbook of Attachment: Theory, Research, and Clinical
Applications*, 2nd ed., Jude Cassidy and Phillip R. Shaver, eds. (New York: Guilford
Press, 2008), 857–79.

11. Images of God the Father can depict either the authoritarian father or the nurturing
daddy. Early stories about God in the first few books of the Old Testament reflect this
first image of the father—determining rules, punishments, and rewards—but by the
time Moses is talking with God by the burning bush, God is kinder and even willing
to negotiate. In the New Testament, Jesus is very Demeter-like in advocating a loving
path, and he calls his Father "Abba," which in Aramaic is the more familiar term for
father, similar to *daddy* in English.

12. Dorothy Dinnerstein, in *The Mermaid and the Minotaur*, postulates that the psychological desire to contain the power of women results from women usually being the primary caretakers of babies. However much babies adore their mothers, they also feel
fear because they are at their mothers' mercy for everything it takes to be comfortable
and survive. Dinnerstein argues that fear of this immensely powerful being stays with
us at an unconscious level throughout life, leading to collusion around keeping wom-
en's power limited. Her solution is to have both men and women involved in infant
care, which would prevent such projection onto women.

13. To my knowledge, no one actually has studied the possible influence of the Eleusinian
Mysteries on the social or political policies of the times, since generally the rites are
studied by mythologists and comparative-religion scholars, not by political and social
historians.

14. With restorative justice, the person who has committed a crime is confronted directly
by the victims. Its goal is to have this be an educational moment that induces a true
change of heart, so that the perpetrator has a chance to experience remorse, make
amends, and then be able to rejoin his or her community and be less likely to engage in
future criminal acts.

Demeter Lesson One: Living a Life of Connected Consciousness

1. Virginia Woolf, *To the Lighthouse* (London: Harcourt Brace Jovanovich, 1955), 147.

2. Cokie Roberts, interview by Renée Montagne, "Book Lauds Women's Role in Founding of Nation," podcast audio, Morning Edition, *NPR*, April 8, 2008. In the NPR
interview, Roberts, author of *Ladies of Liberty: The Women Who Shaped Our Country*, discusses how, in the early years of the nation, women, who had no political or economic
rights, nevertheless were involved in the country's politics. For example, Dolley Madison, the wife of President James Madison, used her hostess skills to bring members of
the rival Federalists and Republicans together at the White House, forcing them to

engage in civil discourse and develop personal relationships so that they would be able to work together better.

3. Frederick Buechner, *Beyond Words: Daily Readings in the ABC's of Faith* (New York: HarperCollins, 2004), 405. This work is a compilation that includes the book in which the quote originally appeared: *Wishful Thinking: A Theological ABC* (San Francisco: Harper & Row Publishers, 1973).

4. Kohlberg's level one (preconventional) is motivated by self-interest, level two (conventional) is defined by social consensus or legality, and level three (postconventional) is governed by universal moral principles.

5. Sally Helgesen, *The Web of Inclusion: A New Architecture for Building Great Organizations* (New York: Doubleday, 1995).

6. Carol S. Pearson, ed., *The Transforming Leader: New Approaches to Leadership for the Twenty-First Century* (San Francisco: Berrett-Koehler Publishers, 2012).

7. Alice H. Eagly and Linda L. Carli, "The Female Leadership Advantage: An Evaluation of the Evidence," *The Leadership Quarterly* 14 (2003): 807–34.

Demeter Lesson Three: Valuing the Generous Heart

1. Claire Cain Miller, "When Women's Goals Hit a Wall of Old Realities," *New York Times,* November 30, 2014, New York edition, BU3.

2. Thomas Merton, *Conjectures of a Guilty Bystander* (Garden City, NY: Doubleday, 1966), 81.

3. Helene P. Foley, ed., *The Homeric Hymn to Demeter: Translation, Commentary, and Interpretive Essays* (Princeton, NJ: Princeton Univ. Press, 1994), 112–18.

4. In Sophocles's play *Antigone,* when Creon refuses to turn over Antigone's brother's remains to be buried properly, she tries everything, including sneaking out at night to attempt to bury him herself. The infuriated Creon sentences her to death by being immured (locked up where she will die of starvation). Creon's son, who loves Antigone, begs him to show mercy, but Creon merely gets furious. Only when the wise Teiresias tells Creon that he has angered the gods and turned the people against him does Creon, in terror, relent. He goes to release her, but she already has hung herself. In sorrow, all the other major figures Creon cares about kill themselves, and the people turn against him. In Euripides's play *Medea,* when Medea's husband, Jason, ignores her and her rights as his wife and takes another wife in her stead, in part for political advantage, her revenge is to kill his new wife and his two sons.

5. Obviously, this interpretation explores a different layer of meaning than Freud's focus on boys' castration fears and girls' penis envy, both of which seem related to societal assumptions that identify power with manhood.

6. Hazel Henderson, "The Ethical Marketplace," lecture, International Conference on Business and Consciousness, Acapulco Princess Resort, Acapulco, Mexico, November 1999.

Demeter Lesson Four: Voting with Your Feet

1. Lisa Belkin, "The Opt-Out Revolution," *New York Times Magazine* 153, no. 52648 (October 26, 2003): 42–86; Pamela Stone and Lisa Ackerly Hernandez, "The All-or-Nothing Workplace: Flexibility Stigma and 'Opting Out' Among Professional-Managerial Women," *Journal of Social Issues* 69, no. 2 (2013): 235–56; Sylvia Ann Hewlett and Carolyn Buck Luce, "Off-Ramps and On-Ramps: Keeping Talented Women on the Road

to Success," *Harvard Business Review* 83, no. 3 (March 2005): 43–54; Sylvia Ann Hewlett, Laura Sherbin, and Diana Forster, "Off-Ramps and On-Ramps Revisited," *Harvard Business Review* 88, no. 6 (June 2010): 30; Lisa A. Mainiero and Sherry E. Sullivan, "Kaleidoscope Careers: An Alternate Explanation for the 'Opt-Out' Revolution," *The Academy of Management Executive* 19, no. 1 (February 2005): 106–23; Michelle Conlin, "The Working-Mom Quandary," *Businessweek* 4037 (2007): 110; and Bernie D. Jones, ed. *Women Who Opt Out: The Debate over Working Mothers and Work-Family Balance* (New York: New York Univ. Press, 2012), vii–viii, http://nyupress.org/webchapters/jones_TOC.pdf.

2. Women have been raising issues about the difficulties of caring for children for decades, yet the hours expected of employees are increasing, not decreasing, and many companies are cutting back on flextime. At the same time, schools continue to operate on schedules set when mothers usually were at home or people lived largely in agricultural communities: summers off, midyear vacations, many early dismissals and holidays, etc. Yet many companies dismiss the problems raised by these conflicting needs and schedules as only "women's issues," which is equivalent to saying, "You are not the norm." Men love their families, too, but often are afraid to speak up about these issues lest they seem insufficiently manly or serious about their work. Sometimes the macho culture of certain fields results in women facing the pain of giving up the work that is their calling. A much-studied example is the shockingly massive brain drain of highly talented women students in the hard sciences who choose not to pursue work in these fields after college. Such female opt-outs report an unwillingness to lead a life conscribed by the built-in macho culture they have experienced in the STEM disciplines.

A study by the Center for Women's Business Research provides data that supports what most executive women know: women left corporate positions because of limited upward mobility (the glass ceiling), a lack of recognition and appreciation for their contributions, and a desire for more flexibility (again related to the need to attend to the care of children and/or elderly parents and other family responsibilities). In so many areas today, women are voting with their feet, withdrawing energy from those places that do not share their values or take their concerns seriously.

According to a Catalyst Research survey cosponsored by the National Association of Women Business Owners (NAWBO), the Committee of 200 (an organization of successful women entrepreneurs and corporate leaders), and Salomon Smith Barney, one-third of women said they were not taken seriously in their prior jobs, and 58 percent reported that nothing would attract them enough to return to the corporate world. Another 24 percent said they might consider going back for more money (likely related to their previously having been paid less than their male colleagues) and 11 percent said greater flexibility might tempt them to return (clearly to be able to meet family responsibilities).

3. Of course, not all of this was pretty, because the emigrants did not know how to respect other peoples and their cultures, nor did they have the sense to learn from them. These new Americans with European values created a new norm and eventually pushed the indigenous population into other locations, sometimes voluntarily, and, sadly, often involuntarily and by force, as also happened with slaves being brought here against their will.

4. John Balzar, "Writer Tom Robbins: A Man of La Conner: Books: He Tilts at the Windmills of American Culture. His Lance Is Humor," *Los Angeles Times* (April 6,

1990), http://articles.latimes.com/1990-04-06/news/vw-773_1_writer-tom-robbins.

5. Tom Robbins, *Even Cowgirls Get the Blues* (New York: Bantam Books, 1990), 244.

Demeter Lesson Five: Standing Up for What You Care About

1. Elizabeth R. Johnson and Katherine A. Tunheim, "Sweden: Pioneering Gender Equality and Advocacy for Women in Contemporary Enterprises and the Home," presentation, International Leadership Association Conference, Montreal, Quebec, October 2013. At the conference session, the presenters and their audience seemed amazed to learn that there is a place in the world where women in leadership positions report not being stressed or driven, while also being happy and prosperous. Sweden's culture is high in caregiver values, which influences the nation's ability to have such policies. That being said, there is significant opposition to these policies. When they were enacted, it was not just to make people happy. The population of Sweden was dwindling and the country needed more workers. The hope was that women would have more children and also stay in the workforce. They did the latter but not the former. Now, the labor shortage is being solved by a large influx of immigrants, but this creates other problems, since they do not necessarily have native Swedes' work ethic or caregiver orientation. Not everyone cares that the policies have increased the national quality of life.

2. According to the 2014 U.N. World Happiness Report, citizens of the happiest countries have longer life expectancies, lower perceptions of corruption, and a high gross domestic product per capita, and they report having more social support, experiencing more generosity, and having more freedom to make life choices. The happiest countries included (1) Denmark, (2) Norway, (3) Switzerland, (4) the Netherlands, (5) Sweden, (6) Canada, and (7) Finland. In the 2015 data that became available as this book was going to press, Sweden slipped to eighth place, but Scandinavian countries still were all in the highest grouping. (The United States was seventeenth in the 2014 report and climbed up to fifteenth place in this new data.)

Zeus Lesson One: Overcoming the Fear That Fuels a Driven Life

1. Oxfam, *Working for the Few: Political Capture and Economic Inequality*, 178 Oxfam Briefing Paper (Oxford, UK: Oxfam GB, January 20, 2014), https://www.oxfam.org/sites/www.oxfam.org/files/bp-working-for-few-political-capture-economic-inequality-200114-en.pdf.

Zeus Lesson Two: Declaring Your Independence

1. A. H. Maslow, "A Theory of Human Motivation," *Psychological Review* 50, no. 4 (1943): 370–96.

2. Of course, illness results from other causes as well, many of them unrelated to how fulfilled we are or are not.

3. James Palmer, "The King's Speech: A Jungian Take," *Jung Journal: Culture and Psyche* 6, no. 2 (Spring 2012): 68–85.

Zeus Lesson Three: Unleashing Your Passion, Focusing Your Actions

1. Contemporary alchemist Catherine MacCoun (in *On Becoming an Alchemist*) explains that the goals of our higher selves often are virtuous, moral, and inspiring, but if they

are not linked to our primal will—i.e., our inner bull—we will not put the needed force behind our resolve to achieve them. Much of the journey to getting what you desire is finding that place where the fire in your belly connects with the brain and also the heart. MacCoun's point is that to perform alchemical magic, both *higher and lower selves* need to be operating.

2. Hillary Rodham Clinton, *Living History* (New York: Simon & Schuster, 2003), 236.

Zeus Lesson Four: Regrouping and Rethinking as You Know More

1. See IPCC, Working Group III, *Climate Change 2014: Summary for Policymakers* (Geneva, Switzerland: IPCC, 2014).

Zeus Lesson Five: Moving from Power Over to Power With

1. In the United States, we are seeing soldiers returning from the wars in Afghanistan and Iraq with a veritable plague of PTSD, as happened with the war in Vietnam, which tells us that we have to rethink war as a test of masculinity.

2. In Greek mythology, Pandora was the first woman made by Prometheus. Zeus, still angry that Prometheus had created humans, gave her a box for a wedding present. When she opened it, all the evils of humankind spilled out; only hope remained.

3. *How Women Lead* website of authors Sharon Hadary and Laura Henderson: http://howsuccessfulwomenlead.com/books/how-woman-lead/key-facts. Their book is *How Women Lead: The 8 Essential Strategies Successful Women Know* (New York: McGraw-Hill, 2012).

4. Sandra L. Bem, "The Measurement of Psychological Androgyny," *Journal of Consulting and Clinical Psychology* 42, no. 2 (1974): 155–62. Bem's research tool used items associated with a wide array of instrumental attributes, such as agency and independence, to assess masculine traits and similarly diverse expressive and relational ones, such as nurturance and warmth, to measure feminine traits. Bem's work evolved over time, which allowed her to also create ways for respondents themselves to rate items as masculine, feminine, or just human. Her reassessment, which shifted the focus from androgyny to enculturation to gender potentially limiting human wholeness as well as limiting the ability to see people as people beyond gender, is explained in *The Lenses of Gender: Transforming the Debate on Sexual Inequality* (New Haven, CT: Yale Univ. Press, 1993). For Bem's updated views on the BSRI, see especially 118–20, 126–27, 154–56, and 206–38.

Persephone and the Way of Transformation

1. To avoid confusion, here and later in this chapter, I will use *Eros* (capitalized) to refer to the god with this name and *eros* (lowercase) will refer to this guidance system.

2. Elaine Pagels, *The Gnostic Gospels* (New York: Vintage Books, 1981). Many now believe Mary Magdalene was Jesus's wife. A good book on this subject is Margaret Starbird's *The Woman with the Alabaster Jar: Mary Magdalen and the Holy Grail* (Rochester, VT: Bear, 1993). Starbird, a staunch Catholic, began her research to disprove what she thought was a distressing idea and ended up convinced it was true. My dear fundamentalist father believed that Jesus had to be married because rabbis at that time were, and because Jesus, he said, came to Earth for the divine to understand all the tribulations of being human. I suspect that he said this after a rare quarrel with my mother.

3. Marion Woodman and Elinor Dickson, *Dancing in the Flames: The Dark Goddess in the Transformation of Consciousness* (Boston: Shambhala Publications, 1996), 8–10.

4. Peter Kingsley, *In the Dark Places of Wisdom* (Inverness, CA: The Golden Sufi Center, 1999), 87–92.

5. Although it has caught on in the popular and scholarly mind that Persephone was raped, no rape is included in the *Homeric Hymns*, and the archaic meaning of the word *rape* includes any *"act of seizing and carrying off by force:* abduction." *American Heritage Dictionary*, 3rd ed. 1996: 1498. What we do know from the story is that Hades's actions would have scared Persephone and were way too sudden for a young girl to process. It appears, however, that this did not alienate her from him permanently. Finally, in the unlikely event that the story told at Eleusis included Persephone being raped, it would make no sense today. Romantic heroines in fiction may get swept away with passion, but they don't typically end up loving a rapist, and neither do real women.

Persephone Lesson One: **Responding to the Call of Eros**

1. See Charlene Spretnak, *Lost Goddesses of Early Greece: A Collection of Pre-Hellenic Myths* (Boston: Beacon Press, 1992), 105–18. Spretnak similarly argues that Persephone would not have been raped. This view also is held by Clarissa Pinkola Estés and argued in *Women Who Run with the Wolves: Myths and Stories of the Wild Woman Archetype* (New York: Ballantine Books, 1992), 263, 412–13. Many versions of the story of Persephone do say that Hades raped her, and in early Orphic versions, some god or another rapes virtually all the goddesses, but these survive only in fragments and do not appear to have been the basis for the Eleusinian rites. It seems to me that rapes in these accounts symbolized being taken over by the power of the gods. Ovid's accounts differ from one another greatly, and he is a bit of a detractor of the gods, including Demeter and Persephone, subtly ridiculing them as he tells their stories.

2. We also can link this to mystical ideas about what happens in the Underworld. Nineteenth-century theosophists, referencing Plato, saw Persephone's, and perhaps also Psyche's, descent as illustrating the soul being purified in death before reincarnation. This happened with the assistance of Hades, who burns away all of the impurities, leaving a person's essence, which is believed always to be beautiful. The same process can be facilitated during one's life through an initiatory rite. Plato and the theosophists agree that this is so that the soul's potential can be realized in the next life or in the next phase of this one.

3. Esther Perel, *Mating in Captivity: Unlocking Erotic Intelligence* (New York: Harper, 2006), 1–18.

4. Deborah Tannen, *You Just Don't Understand: Women and Men in Conversation* (New York: Ballantine Books, 1990), 77, 179.

Persephone Lesson Two: **Claiming Your Love Rights**

1. Meg Cabot, *Abandon* (New York: Point, 2011); Meg Cabot, *Underworld* (New York: Point, 2012); and Meg Cabot, *Awaken* (New York: Point, 2013).

2. I'm reminded by this that in ancient Minoan Crete, many of the most sacred rituals happened in caves and often required a long trek down into the earth. When I was in Colombia for a Fulbright program, wonderful people there took me to a salt mine where the miners had created an altar made of salt and a statue of the Black Madonna.

Some of the ancient awareness of the earth as sacred clearly remains in human intuitive life today.

Persephone Lesson Three: Doing Life a Simpler Way

1. His story, of course, uses their Latin names, so it is called *Cupid and Psyche*.
2. See Louann Brizendine, *The Female Brain* (New York: Broadway Books, 2006), which describes a range of hormonal and brain structures that affect gender.
3. Stieg Larsson's The Girl with the Dragon Tattoo trilogy is a Persephone-like story corresponding to its more tragic versions where Persephone is raped. Lisbeth Salander, the central character of the novels, is abused horribly through most of her life—by her father, in a mental hospital, and in the criminal underworld. Her trials have many elements of the Persephone story, and to a lesser extent, the Psyche story; her eventual triumph results from her intelligence, intuition, and courage, along with the fact that others help her, even though, in many parts of the series, Lisbeth does not know this and feels very much alone.

Persephone Lesson Four: Making Choices to Realize Your Destiny

1. Scholar Jessie Weston's seminal book *From Ritual to Romance* traces the origins of romantic love and chivalry depicted in the Arthurian legends back to the ancient fertility religions, whose traditions the Eleusinian Mysteries preserved.
2. Stephanie can be seen as living the archetypal plot from the Arthurian legends, where she is in the role of Guinevere, King Arthur's wife, who is kidnapped repeatedly by evil knights and then rescued by Lancelot. However, when she and Lancelot act on their erotic feelings for one another, it spells the end of Camelot. For a variety of reasons, professional codes in psychology and education forbid psychologists and educators from getting romantically—or sexually—involved with their clients and students. Because Ranger refuses to divulge much about himself, Stephanie projects her idealized self onto him—the self she is becoming with his help. Since neither has an official power relationship over the other, their flirting is not a violation of some professional code, nor does it undercut Ranger's potency as a role model for Stephanie.
3. If the bride declined the food, new discussions would have to be held about what would be acceptable to her. Wives also had rights to leave their husbands with the assistance of their families, taking their dowries with them, although a male relative would manage the money. However, the dowries still had to be used for the woman's benefit. In addition, anyone in ancient Greece would know that if you do not want to stay in the Underworld, you had better not eat anything while there.
4. See Foley, *Homeric Hymn to Demeter*, 56–57 in the commentary on the translation, and 104–11.
5. Richard Tarnas (author of *The Passion of the Western Mind*), in discussion with the author, Assisi Institute Conference, spring 2013.

Persephone Lesson Five: Experiencing Radical Belonging

1. The belief that the hieros gamos was practiced in the Mysteries was based on charges by a Roman Christian cleric who was scandalized by the thought that there was such a sexually explicit reference at Eleusis. His reaction reflected the antisex bias of the

church that began in Rome in reaction to what was a culture desperately needing to rein in sensual self-indulgence in many spheres (think of the seven deadly sins, all of which were widely in evidence). The Greeks were not antisex, but they did expect wives to be faithful to their husbands, at least partly for practical reasons related to patriarchal inheritance practices.

2. Brian Swimme, *The Universe Is a Green Dragon: A Cosmic Creation Story* (Santa Fe, NM: Bear, 1985), 45.

3. Martin P. Nilsson, "The Religion of Eleusis," *Greek Popular Religion* (New York: Columbia Univ. Press, 1940), 42–60. Nilsson summarizes scholarship that connects Persephone with the hieros gamos, with the birth of the divine child, the origins of Persephone in Crete (including the Minotaur and the labyrinth), and the meaning of eating the pomegranate seeds. See also Jean Houston, who, in *The Search for the Beloved: Journeys in Mythology and Sacred Psychology,* describes the marriage of Persephone and Hades as a hieros gamos that unifies the world of light with the realm of shadow, which happens when you come to value your previously disowned parts, helping you to become more whole.

4. Elisabet Sahtouris, *Gaia: The Human Journey from Chaos to Cosmos* (New York: Pocket Books, 1989), 159–62.

5. Sahtouris, *Gaia*, 20.

6. Sahtouris, *Gaia*, 26–27.

7. Margaret J. Wheatley, *Leadership and the New Science: Discovering Order in a Chaotic World* (San Francisco: Berrett-Koehler Publishers, 1994), 100.

8. I use the men's last names and the women's first names only because that is how they are addressed in the movie, and otherwise it would be difficult to tell the two daughters from their fathers, since both have their paternal last name.

9. Dave Itzkoff, "A World Flamed into Life," an interview with Barbara Ehrenreich, *New York Times Magazine* (March 30, 2014): 18.

Dionysus and the Gift of Joy

1. Angeles Arrien, *The Four-Fold Way: Walking the Paths of the Warrior, Teacher, Healer, and Visionary* (San Francisco: HarperOne, 1993,) 41. Arrien also says that dancing helps us retrieve parts of the self that have been lost or forgotten.

2. Estés, *Women Who Run with the Wolves,* 11.

3. Robert Bly, *Iron John: A Book About Men* (Reading, MA: Addison-Wesley, 1990).

4. The danger in this prescription is that, in practice, men can feel a continuing need to devalue women and feminine qualities in themselves in order to stop wanting to please us by acting, as Bly fears, in softer, more feminine ways. Yet just as women are more likely to claim their instinctual power if men are strong enough not to need to devalue us to bolster their own egos, men's ability to assert their power depends upon heroines claiming ours as well. If we do not, men in their power will scare and intimidate us. When we affirm our feminine instinctual power and free ourselves from the desire to leave our truths to please men (father, boyfriend, men in power in the world), it is important that we do not then disdain the masculine. Otherwise, we too will be unable to claim our wholeness or to offer ourselves as a worthy lover to a man (or another woman).

5. To read this poem or find out how to hear Mary Oliver read it, go to http://www.brain pickings.org/2014/09/24/mary-oliver-reads-wild-geese/.

6. His rites were associated with drunkenness in Rome, but not in Greece. However, various forms of debauchery are associated with the negative side of the overarching archetype.

7. "I am the vine, ye are the branches: He that abideth in me, and I in him, the same bringeth forth much fruit: for without me ye can do nothing" (John 15:5, KJV).

8. Details about what happened in the rites have been pieced together by various scholars, often from brief references, and the experts do not all agree on every detail. For example, we know there was a statue of Dionysus and many musical instruments, but we have no details about them. What I've added—through phrases that indicate they are my imaginings of what might have been—signals that these details are not actually included in the sources I have referenced or otherwise utilized. As far as I know, those specifics are not actually known.

9. Clement of Alexandria, *Protreptikos* II, 18, *Clement of Alexandria with an English Translation by G. W. Butterworth* (Cambridge: Harvard Univ. Press, 1953), 43.

10. Daryl Sharp and C. G. Jung, *Jung Lexicon: A Primer of Terms and Concepts* (Toronto: Inner City Books, 1991), 109–10.

Dionysus Lesson One: Realizing the Eleusinian Promise

1. Ariadne was the sister of the Minotaur and helped the hero Theseus find his way through the labyrinth to kill the creature, who was demanding human sacrifices. Upon his return, they took off by ship together, but Theseus abandoned her on the first island where they stopped. Dionysus, who became known for disdaining hero types as well as embodying an alternative to them, rescued her. However, it is important that one of his ascribed birth mothers was a mortal, as was Ariadne, who eventually died. Dionysus descended into the Underworld and brought both back, demonstrating the same ability to traverse the worlds possessed by Persephone, Hermes, and Hekate.

2. Some experts believe that the initiates not only danced in the public procession, but that together they also would have danced a well-choreographed version of the most secret story while cloistered in its most private rites. In addition, there is ample evidence that dance is a very ancient female shamanistic practice that involved such trance dancing, which had a function important to the feminine side of spirituality that went way beyond catharsis. While the ancient male shamanistic path was more individualistic, women danced together to foster an outcome, in the process ushering in the archetypal energy that was needed for the time. Perhaps as a way of preserving a legacy that worked for them, women flocked to the Dionysian rites as they did to the Eleusinian ones, both of which included dance.

3. Others, researching this, say that the maenads, his followers, likely did not drink wine at all, except possibly as a kind of communion ritual along with meat cooked in milk, although some report that one Dionysian ritual involving men used wine in larger quantities. In addition, there are stories, like the one about Jesus at the wedding feast, where Dionysus turned water into wine. We can see alcoholism, as well as any kind of drug addiction, as a kind of madness that substitutes for the divine ecstatic or at least connection with one's spiritual nature, as is widely understood in twelve-step programs, although those don't include wild and raucous dancing.

4. Mary Wilson, "Dionysus: Transforming Presence in Psyche and the World" (master's thesis, Pacifica Graduate Institute, 1989), 40. She provides one of the best summaries I've found of the literature on Dionysus, and her thesis is enjoyable to read.

5. James Hillman, *Re-Visioning Psychology* (New York: Harper & Row Publishers, 1975), 104. As with many quotations that often are paraphrased by its author or others in oral communication, the original published quote was longer, but its length is helpful to provide context about what he meant. "Archetypal psychology can put its idea of psychopathology into a series of nutshells, one inside the other: within the affliction is a complex, within the complex an archetype, which in turn refers to a god. Afflictions point to gods; gods reach us through afflictions. . . . Zeus no longer rules Olympus but rather the solar plexus, and produces curious specimens for the doctor's consulting room." Then Hillman refers to Jung saying that the gods or archetypes now are present in diseases. Hillman then continues, "Gods, as in Greek tragedy, force themselves symptomatically into awareness. Our pathologizing is their work, a divine process working in the human soul. By reverting the pathology to the God, we recognize the divinity of pathology and give the God his due."

6. Bonnie J. Horrigan, "Meaning as a Healing Agent," *Explore: The Journal of Science and Healing* 8, no. 6 (November/December 2012): 323–25.

7. This healing sanctuary was founded in Santa Barbara but now does most of its work in Mexico. See http://santabarbarahealingsanctuary.com/the-malinalco-healing -sanctuary/.

Dionysus Lesson Two: Celebrating Life's Great Beauty

1. The public forms of ancient Greek religions lacked a sense of the great mystery of a divine energy that was both beyond this world (transcendent) and in it (immanent), leaving us with archetypes that primarily embody either natural processes (as with the Titans) or human attributes (as with the Olympian gods). The recognition of something divine beyond all that has been the contribution of monotheistic religions. Because theologians saw the world through the culture's patriarchal lens, the divine was made understandable by presenting it in the Zeus-like image of a white male king or father, defined according to how kings and fathers were experienced in different regions and eras. Over time, some versions of monotheistic religions focused on transcendence, denying immanence and seeing evil in the body, sexuality, and the material world. Perhaps this was because life in many historical periods was—as it remains in too many places today—"nasty, brutish, and short," as the philosopher Thomas Hobbes described it in *Leviathan*. Therefore, people were encouraged to sacrifice in this life in order to gain happiness in the next, or to find relief from suffering by not desiring anything. Hobbes saw this sad condition as the "state of nature." But we might recognize that not-so-enlightened human causes also play a part in creating dystopian realities.

2. Episcopal priest and theologian Matthew Fox (in *Original Blessing* and other related books) traces the path of joy and beauty back to Jesus through various Christian mystics and back through mystics in many other religions, both modern and indigenous. He labels this tradition "panentheism," which, like pantheism, sees the divine in nature but with an existence beyond it as well. He notes that many religions engage the body in ritual dancing or other movements, as with swaying while praying in Judaism or ecstatic whirling in Sufism, so he incorporates body prayer in his form of modern worship. This and similar contemporary ecumenical efforts are reclaiming the body, sexuality, and the natural world as sacred while recognizing that the divine is a potential in men and women, all races and ethnic groups, and people along the continuum of sexual orientation. In *The Coming of the Cosmic Christ,* he differentiates "the Christ"

from the person of Jesus and references definitions of it as a state of consciousness, an archetype that is available to all but that Jesus completely embodied. His thinking makes it possible for Christians to remain in their faith even as they respect and learn from truths in other religions.

3. Many reviewers saw this movie as less hopeful than I did—but then, none of them caught the connection with Dionysus. The following review interprets the film similar to the way I have: "'It's Just a Trick': Dissecting the Final Scene of 'The Great Beauty,'" *Viva Italian Movies,* last modified June 20, 2014, http://www.vivaitalianmovies.com /post/80850166067/great-beauty-final-scene.

Dionysus Lesson Three: Dancing Collective Joy

1. Clare Farnsworth, "Seahawks Flipped the Switch to Power 9-1 Finish," Seahawks News post, December 29, 2014, http://www.seahawks.com/news/2014/12/29/seahawks -flipped-switch-power-9-1-finish.

2. "Lord of the Dance," nineteenth-century Shaker melody adapted by Sydney Carter (Carol Stream, IL: Hope Publishing, 1963).

3. See www.originalblessing.ning.com.

4. The practice of Nia, developed by Debbie Rosas and Carlos Rosas (*The Nia Technique*), is helpful for liberating the energy of all parts of the body in ways that are designed to work with its anatomy. Nia also fosters the practice of dancing through life in touch with the pleasure of sensation. See www.nianow.com.

5. The dance form of the 5Rhythms, based on the work of Gabrielle Roth (*Sweat Your Prayers*), provides a Dionysian-like practice of moving with controlled wildness, dancing first to a flowing, then staccato, then chaotic beat and then, when catharsis is reached, to music that becomes lyrical and then evokes meditative stillness. This progression helps to free the body from painful emotions that are held in it and that restrict energy flow. See www.5rhythms.com.

6. The vertical is longer than the horizontal on the traditional Christian cross, as it is on the Egyptian and Coptic ankh. On the Greek cross, the Celtic cross, and the modern Red Cross, the vertical and horizontal are equal in length, which makes the horizontal focus on care for others equivalent to the vertical focus on love for God.

7. To see flash-mob events all over the United States, go to www.flashmobamerica.com /videos.

8. Barbara Ehrenreich, *Dancing in the Streets: A History of Collective Joy* (New York: Metro-politan Books, 2007), 26.

Dionysus Lesson Four: Directing Your Inner Theater Company

1. Jane Wagner, *The Search for Signs of Intelligent Life in the Universe* (New York: Harper & Row Publishers, 1986), 18.

2. For more on subpersonalities, see John Rowan, *Subpersonalities: The People Inside Us* (London: Routledge, 1990).

The Power of Story to Transform Your Life

1. Changes in gender roles, of course, did not cause this shift; they just happened around the same time.

2. Margaret Mark and Carol S. Pearson, *The Hero and the Outlaw: Building Extraordinary Brands Through the Power of Archetypes* (New York: McGraw-Hill, 2001).
3. Steven M. Southwick and Dennis S. Charney, "Ready for Anything," *Scientific American Mind* 24, no. 3 (July/August 2013): 32–41. This article is based on *Resilience: The Science of Mastering Life's Greatest Challenges* by the same authors.
4. Jennifer G. Bohanek, et al., "Family Narrative Interaction and Children's Sense of Self," *Family Process* 45, no. 1 (March 2006): 39–54. This study was conducted by a team from Emory University. See also Amber Lazarus, "Relationships Among Indicators of Child and Family Resilience and Adjustment Following the September 11, 2001 Tragedy" (working paper no. 36, The Emory Center for Myth and Ritual in American Life, 2004).
5. Zachary Green, "Unleashing Possibilities: Leadership and the Third Space," *The Transforming Leader: New Approaches to Leadership for the Twenty-First Century,* Carol S. Pearson, ed. (San Francisco: Berrett-Koehler Publishers, 2012), 211–19.

Reading Group Discussion Guide

Though each archetypal lesson in *Persephone Rising* ends with an application exercise, and each part with a capstone exercise—both designed to engage you deeply—the brief questions that follow are meant to serve as suggested sparks to light the fire of larger, longer group discussions.

1. How is the telling of the Demeter and Persephone myth in *Persephone Rising* different from any earlier versions of the story you may have encountered? What elements of Pearson's telling of the story stand out to you and why?
2. After reading the whole book, how is your view of the narrative, now including the perspective of all four major characters, different from when you first read the opening story?
3. Who do you know who embodies elements of the Demeter, Persephone, Zeus, or Dionysus archetype? Where do you see any of these archetypes in yourself?
4. Think of sometime you or someone you know has felt abducted by fate, as Persephone is in this myth. What would, or did, it

mean to you or them to make a creative choice, similar to her eating the pomegranate seeds? How did this work out or how do you think it would?

5. Demeter creates a famine to get her daughter back. What loss or disregard might lead someone today to go to such extreme ends? What is the limit of how far you believe it would be okay to go to protect someone or something you love?

6. Zeus moves from feeling powerless as an exiled child to becoming the ruler of the gods. What is the territory you would control if you let yourself express your full power?

7. Thinking of Dionysus's dancing as a metaphor for feeling centered in yourself and in relationship with others and the world, when have you most memorably experienced such a feeling?

8. What do you imagine when you imagine "having it all"? How might activating the different archetypes of Demeter, Zeus, Persephone, and Dionysus empower you to realize this vision?

9. When was a recent time—in conversation, listening to the radio, reading a paper, watching the news, skimming a magazine—when you felt caught up in a narrative that on reflection didn't feel right to you? And how might you, or did you, employ your sword and shield to reassess that narrative?

10. What is an example of a grail story you tell about your life, which, in telling, gives you a sense of purpose and reflects the most genuine aspects of your identity?

11. Do you have an inner script that could use a narrative overhaul? How might you use your magic wand to rewrite that plotline?

12. If you were to create your own Mystery tradition today, what would you include in the ritual and why? How might you incorporate the essence of Mystery traditions in modern life?

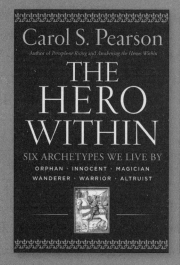
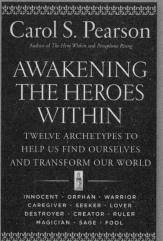